The Energy Transition in Transport

Towards Sustainable Mobility

Francisco José Hurtado Mayén

Content

Introduction: The Need for a Transition in Transportation 7

    The environmental impact of transport 8

    The evolution towards sustainable transport 9

    Electric Vehicles: A Revolution in the Making 10

    Public transport and shared mobility 11

    Public Policy: The Government's Role in the Transition 12

    Case Studies: Success Stories 13

    Challenges and opportunities in the transition 15

    Conclusion 16

Chapter 1: History and Evolution of Electric Vehicles 19

    First Electric Vehicles: The Pioneers of the Nineteenth Century 20

    Internal combustion engine and the oil crisis of the 70s 21

    Challenges and barriers to EV adoption 26

    The future of electric vehicles 27

    Conclusion 29

Chapter 2: Innovations in public transport and mobility 31

    Electric public transport: Electric buses and trams 32

    Shared mobility: Carsharing, bikesharing and ridesharing ... 33

    Autonomous vehicles and their integration 35

    Impact of shared mobility 36

    Policy Policies in Sustainable Technologies 38

    Future Perspectives: Innovations and Emerging Trends 41

    Conclusion 43

Chapter 3: Policy Impact on Sustainable Technologies 47

    Subsidies and economic incentives 48

Regulations and standards 50

Infrastructure investments 54

Public-private partnerships 56

Impact of public policies on sustainable technologies 59

Challenges and lessons learned 61

Innovations and trends in public policies 64

Conclusion 67

Chapter 4: Case Studies and Future Perspectives 69

Case Studies: Innovations and Successes in Sustainable Mobility 70

Future Outlook: Emerging Trends and Opportunities 75

Conclusion 78

Chapter 5: Technological Innovation in Sustainable Mobility 81

New generation electric vehicles 82

Wireless and fast charging 84

Telematics and connected vehicles 86

Emission reduction technologies 88

Conclusion 91

Chapter 6: Sustainable Mobility in Smart Cities 95

Definition and characteristics of a smart city 96

Urban mobility and smart transport 98

Traffic Management & Urban Planning 100

Sustainable mobility technologies in smart cities 103

Impact of mobility on smart cities 106

Challenges and barriers to implementation 108

Future of mobility in smart cities 111

Conclusion 114

Chapter 7: Citizenship and Education in Sustainable Mobility 115

    Awareness campaigns and public education 116

    Community Participation in Transportation Planning 119

    Educational programs and training.. 122

    Impact of citizen participation and education................................ 125

    Challenges and barriers to citizen participation and education 128

    Future of participation and education in sustainable mobility 131

    Conclusion... 134

Chapter 8: Economic Benefits of Sustainable Mobility......................... 137

    Reduced operating costs.. 138

    Economic growth and job creation 140

    Public Health Savings 144

    Economic incentives and return on investment.............................. 147

    Examples of positive economic impact on sustainable mobility 150

    Economic challenges of sustainable mobility.................................. 153

    Future prospects and emerging opportunities 156

    Conclusion... 160

Chapter 9: Sustainable Mobility in Rural Areas 161

    Challenges specific to rural regions... 162

    Solutions adapted for rural mobility... 165

    Support policies and programmes 167

    Success stories in rural mobility 170

    Implementation strategies and best practices................................ 173

    Future prospects and emerging opportunities 175

    Conclusion... 178

Conclusion: Towards a Sustainable Mobility Future 181

    Recap of key learnings.. 182

    Emerging challenges and opportunities.. 187

Future prospects for sustainable mobility 189

Vision for a sustainable future 191

Conclusion 193

Introduction: The Need for a Transition in Transportation

Transport has been and continues to be one of the fundamental pillars of modern society. From the first horse-drawn carriages to airplanes that cross continents in a matter of hours, advances in transportation have radically transformed the way we live, work, and relate to each other. However, this impressive progress has come at a significant cost to the environment and public health. Today, the transportation sector is responsible for a large portion of greenhouse gas emissions and air pollution in our cities.

In the face of the growing climate crisis and growing air quality concerns, the need for a transition to more sustainable modes of transport has become imperative. This book not only addresses this urgent need, but also aims to inspire and motivate governments, businesses and citizens to take concrete and effective action.

In these pages, we will explore how advanced technologies, effective public policies, citizen participation, and economic benefits can converge to create a more sustainable and equitable transportation system. The success of sustainable mobility depends on our ability to work together, innovate and maintain a strong commitment to sustainability.

With the right approach, we can build a transportation system that not only meets our mobility needs, but also protects our planet and improves the quality of life for all. We hope that the lessons and strategies presented in this book will serve as a guide and inspiration to build a cleaner, more efficient and fairer mobility future for generations to come. With determination and collaboration, we can move towards a sustainable mobility future that benefits everyone and ensures a healthier and more prosperous planet.

The environmental impact of transport

Transportation, as we know it, relies heavily on fossil fuels. Cars, trucks, planes, and ships mainly use gasoline, diesel, and kerosene, all of which are derived from petroleum. The combustion of these fuels produces carbon dioxide (CO2), the main GHG, along with other harmful pollutants such as nitrogen oxide (NOx) and fine particulate matter (PM2.5). These emissions not only contribute to climate change, but also have serious repercussions for human health.

Climate change is perhaps the most urgent challenge of our time. The accumulation of GHGs in the atmosphere is causing global temperatures to rise, which in turn leads to extreme weather events such as more intense hurricanes, prolonged droughts, and devastating wildfires. Emissions from the transport sector account for about a quarter of global CO2 emissions, and reducing these emissions is crucial to mitigating the effects of climate change.

Air pollution is another critical issue associated with transportation. According to the World Health Organization (WHO), air pollution is responsible for millions of premature deaths each year. Air pollutants can cause respiratory and cardiovascular diseases, lung cancer, and other serious health problems. Cities with high levels of traffic often suffer the worst levels of air pollution, disproportionately affecting low-income and vulnerable communities.

The evolution towards sustainable transport

The transition to sustainable transport is not just a matter of reducing emissions and pollution. It's also about building a transportation system that is equitable, accessible, and resilient. Sustainable transport must be able to meet the mobility needs of all people, regardless of their location or economic situation, and must be prepared to meet future challenges, from urban growth to the impacts of climate change.

In recent decades, there has been significant progress in developing technologies and strategies to make transportation more sustainable. Electric vehicles (EVs) have gone from being a technological curiosity to a viable and increasingly popular option. Innovations in public transportation, such as electric buses and rapid transit systems, are helping to reduce reliance on private cars. Shared mobility solutions, such as carsharing and bikesharing, are transforming the way people get around in cities.

Electric Vehicles: A Revolution in the Making

Electric vehicles are one of the most promising technologies for reducing transport emissions. Unlike vehicles with internal combustion engines, EVs do not emit CO_2 or other pollutants during their operation. In addition, EVs can be powered by renewable energy sources, further reducing their environmental impact.

The resurgence of EVs over the past two decades has been driven by several factors. First, there have been significant advances in battery technology, especially lithium-ion batteries. These batteries are lighter, have a higher energy storage capacity, and are more durable than previous generation batteries. As a result, modern EVs have a much longer battery life and are more practical for everyday use.

Second, governments around the world have implemented policies to promote EV adoption. These policies include direct subsidies for the purchase of EVs, tax incentives, restrictions on polluting vehicles, and the development of charging infrastructure. These measures have helped to reduce the total cost of ownership of EVs and increase their attractiveness to consumers.

In addition, the automotive industry has made massive investments in EV development and production. Companies such as Tesla, Nissan, General Motors, and many others are bringing an ever-expanding range of electric models to the market. This competition has accelerated innovation and contributed to lower EV prices.

Public transport and shared mobility

While EVs offer a solution for individual commuting, public transportation and shared mobility solutions are essential to reduce the overall reliance on private cars. An efficient and accessible public transport system can move large numbers of people much more sustainably than private vehicles.

Electric buses and rapid transit systems (such as light rail and subways) are examples of how public transport can adapt to be more sustainable. Electric buses, in particular, are gaining popularity in many cities due to their lower operating cost and lower environmental impact compared to traditional diesel buses. In addition, rapid transit systems can transport large numbers of people efficiently and with low emissions.

Shared mobility solutions, such as carsharing, bikesharing, and ridesharing, are also playing an important role in the transition to more sustainable transportation. These options allow people to access a vehicle or bicycle when they need one, without having to own one. Not only does this reduce the number of vehicles on the roads, but it also encourages the use of more sustainable modes of transport, such as walking and cycling.

Public Policy: The Government's Role in the Transition

Public policies are crucial to facilitate and accelerate the transition to sustainable transport. Governments can implement a variety of measures to promote the use of sustainable transport technologies and practices. These measures include economic incentives, regulations, investments in infrastructure and awareness campaigns.

Economic incentives, such as EV purchase subsidies and tax breaks, can make sustainable vehicles and technologies more affordable for consumers. Regulations, such as emission standards and low emission zones, can force transport manufacturers and operators to reduce their environmental impact.

Investments in infrastructure, such as the construction of EV charging stations and the expansion of public transport networks, are essential to support the use of sustainable technologies. Without adequate infrastructure, consumers may find it difficult or inconvenient to adopt new technologies.

Finally, awareness campaigns can help educate the public about the benefits of sustainable transport and encourage changes in behaviour. These campaigns can include information on the environmental impacts of transport, as well as practical advice on reducing car use and opting for more sustainable modes of transport.

Case Studies: Success Stories

Around the world, there are numerous examples of cities and countries leading the way in the transition to sustainable transport. These case studies offer valuable lessons and demonstrate that, with the right mix of policies, technology and political will, meaningful change is possible.

Norway: Norway is a global leader in EV adoption, with a significant percentage of its population driving electric cars. This has been made possible by generous government incentives, such as tax exemptions, tolls, and free parking for EVs. In addition, Norway has invested heavily in charging infrastructure, making it easier to use EVs on a daily basis.

China: China has made great strides in electrifying public transportation, especially in its buses. Cities like Shenzhen have fully electrified their bus fleets, significantly reducing emissions and improving air quality. The Chinese government has supported these efforts with subsidy policies and strict emissions regulations.

United States: In the United States, several cities and states have implemented successful policies to promote sustainable transportation. California, for example, has set ambitious targets for emissions reductions and implemented a number of incentives for EVs and other clean technologies. In addition, cities like New York and Seattle are investing in expanding and improving their public transportation systems.

Latin America: In Latin America, cities such as Bogotá and Santiago have made significant strides in promoting public transport and sustainable mobility. Bogota is known for its bus rapid transit (BRT) system,

TransMilenio, which has significantly improved mobility and reduced emissions. Santiago, for its part, has implemented a plan to electrify its bus fleet and has improved its metro network.

Challenges and opportunities in the transition

While there are many examples of success, the transition to sustainable transport faces a number of challenges. One of the main challenges is the upfront cost of new technologies. Although EVs and electric buses may be cheaper to operate in the long run, their initial cost is still high compared to traditional alternatives. This can be a barrier for consumers and cities with limited resources.

Another challenge is infrastructure. EV adoption requires an extensive and reliable network of charging stations, and the expansion of public transportation requires significant investments in new lines and vehicles. In many places, existing infrastructure is not prepared to support these new technologies, and the necessary investments can be costly and time-consuming.

In addition, there are challenges related to acceptance and behavior change. People may be reluctant to change their transportation habits, especially if they are used to the convenience of the private car. Promoting the use of public transport and shared mobility solutions requires cultural and behavioural changes, as well as improvements in the quality and accessibility of these services.

Despite these challenges, the transition to sustainable transport also presents many opportunities. Technological innovations continue to reduce costs and improve the performance of sustainable technologies. Public policies can play a crucial role in overcoming barriers and facilitating change. And, perhaps most importantly, the transition to sustainable transport can offer significant benefits for the environment, public health and overall quality of life.

Conclusion

The need for a transition to sustainable transport is clear. The environmental and health impacts of the current fossil-fuel-based transport system are unsustainable, and the climate crisis requires urgent action. Fortunately, there are many technologies and strategies available that can help us achieve this transition.

Electric vehicles, electric public transport, and shared mobility solutions offer viable paths to reduce emissions and improve transport sustainability. Public policies and infrastructure investments are essential to support these technologies and facilitate their adoption. And case studies from around the world show that with the right mix of policies, technology, and political will, meaningful change is possible.

This book will explore these topics in depth, providing a comprehensive guide on the transition to sustainable transport. We will look at the history and evolution of EVs, innovations in public transport and shared mobility, effective public policies and inspiring case studies. Through this exploration, we hope to provide a clear and compelling vision of how we can build a more sustainable and equitable transportation system for the future.

Chapter 1: History and Evolution of Electric Vehicles

Electric vehicles (EVs) are revolutionizing the way we understand transportation and energy. Unlike vehicles with internal combustion engines, which burn fossil fuels to generate power, EVs run on electricity stored in batteries. This change not only reduces greenhouse gas emissions, but also offers the possibility of using renewable energy sources to power transport. To understand the impact and potential of EVs, it is essential to explore their history, the technological advances that have enabled their resurgence, and the challenges they still face.

First Electric Vehicles: The Pioneers of the Nineteenth Century

The history of electric vehicles dates back to the nineteenth century, a time of great innovation in transportation technology. The first prototypes of electric vehicles appeared almost simultaneously with the first internal combustion cars. In the late 19th and early 20th centuries, EVs were considered serious competitors to gasoline and steam cars.

One of the first milestones in the history of EVs was the work of British inventor Thomas Parker, who built an electric vehicle in 1884 using high-capacity rechargeable batteries. At the same time, in the United States, inventors such as William Morrison of Des Moines, Iowa, were also developing their own electric vehicles. Morrison's car, built in 1891, could carry up to six passengers and reach a top speed of 23 km/h.

At the end of the nineteenth century, EVs had certain advantages over their competitors. They were quieter and easier to operate than gasoline-powered vehicles, which required manual engine starting, a physically demanding task. In addition, EVs did not emit smoke or unpleasant odors, which made them more attractive for urban use. In 1899 and 1900, EVs accounted for about one-third of all vehicles on United States roads.

Internal combustion engine and the oil crisis of the 70s

Despite their early advantages, EVs began to lose ground to internal combustion cars in the early decades of the 20th century. Several reasons contributed to this decline. First, internal combustion engines improved rapidly in terms of performance and reliability. The invention of the electric starter motor by Charles Kettering in 1912 eliminated the need to manually start cars, making gasoline vehicles easier to use.

In addition, the mass production of gasoline cars, initiated by Henry Ford with the Model T, significantly reduced production costs and, therefore, the selling price. Gasoline-powered cars became more accessible to the general public, while EVs, which were still expensive to produce, were unable to compete in terms of price.

Another crucial factor was infrastructure. The network of petrol filling stations expanded rapidly, while the infrastructure for EV charging remained limited. The combination of these factors led to an almost total dominance of internal combustion vehicles in the mid-20th century.

However, the oil crisis of the 1970s sparked renewed interest in EVs. High oil prices and concerns about energy dependence drove research and development of alternatives to fossil fuels. During this period, several automakers and governments began to explore the potential of EVs again. Although advances were limited and EVs failed to achieve mass adoption at the time, the groundwork was laid for future developments.

Key Technologies: Batteries, Motors & Energy Management

The resurgence of electric vehicles (EVs) over the past two decades has been made possible by a number of technological advances, particularly in the field of batteries, electric motors, and energy management systems. Lithium-ion batteries have been a crucial factor in the EV renaissance. Invented in the 1980s, these batteries offer higher energy density and longer lifespan compared to the lead-acid batteries used in early EVs. The energy density of lithium-ion batteries allows more energy to be stored in a smaller space, resulting in vehicles with longer ranges. In addition, lithium-ion battery costs have fallen dramatically in recent years, thanks to improvements in manufacturing processes and increased production on a global scale. This cost reduction has been instrumental in making EVs more accessible to the average consumer.

The electric motors that power EVs have also undergone significant advances. Permanent magnet motors, which use rare earth magnets, offer high efficiency and high performance compared to traditional electric motors. These engines are more compact and lighter, which contributes to improving the energy efficiency of vehicles. In addition, advances in power electronics have allowed for more precise control of electric motors, improving the vehicle's responsiveness and overall performance. Energy regeneration during braking, a feature present in most modern EVs, allows some of the kinetic energy to be recovered and stored in the battery, further increasing the vehicle's efficiency.

Energy management systems are essential for optimizing performance and battery life in EVs. These systems continuously monitor the battery health, manage charging and discharging, and ensure that the electric motor operates in optimal conditions. Thermal management technology also plays a crucial role, keeping batteries and motors at an ideal temperature to maximize their efficiency and longevity. In addition, energy management systems in modern EVs are integrated with advanced software that can dynamically adjust energy consumption based on driving conditions, driver style, and other factors. This artificial intelligence and machine learning are starting to play an increasingly important role in optimizing EV performance.

Adoption and Emerging Technologies

Today, electric vehicles (EVs) are on an upward trajectory, with significant growth in sales and adoption worldwide. Several factors are driving this trend, including the technological advancements mentioned above, favorable government policies, and growing consumer interest in more sustainable transportation solutions.

In recent years, EV sales have increased exponentially. In 2020, despite the COVID-19 pandemic, global EV sales exceeded 3 million units, a significant increase compared to previous years. This growth has been particularly notable in markets such as China, Europe and, to a lesser extent, the United States.

Governments around the world are implementing policies to encourage EV adoption. These policies include direct subsidies for EV purchases, tax incentives, restrictions on internal combustion vehicles, and the development of charging infrastructure. For example, Norway has implemented a series of incentives that have led to more than half of the new cars sold in the country being electric.

Charging infrastructure is a critical component to EV success. In response to increased demand, charging station networks are rapidly expanding. Companies like Tesla have developed their own supercharger networks, while other manufacturers and energy providers are investing in building public and private charging stations.

In addition to existing technologies, new innovations are being developed that could further transform the EV market. Solid-state batteries, for example, promise higher energy density and faster charging times compared to current lithium-ion batteries. These batteries use a solid electrolyte instead of a liquid one, which improves safety and efficiency.

Another area of development is wireless charging, which would allow EVs to charge without the need to physically connect to a charging station. This technology, still in experimental phases, could make EV charging more convenient and accessible.

Autonomous vehicles are also emerging as a complementary technology to EVs. These vehicles, which use sensors, artificial intelligence and advanced connectivity to navigate without human intervention, have the potential to improve transport efficiency and further reduce emissions. Electric autonomous vehicles could optimize driving routes, minimize energy consumption, and facilitate the integration of renewable energy into the transportation system.

Challenges and barriers to EV adoption

Despite significant progress, the mass adoption of electric vehicles (EVs) still faces several challenges. These challenges must be addressed to ensure EVs can deliver on their promise to transform transport and reduce emissions.

Although battery costs have declined, EVs are still more expensive than comparable internal combustion vehicles. This higher upfront cost can be a barrier for many consumers, especially in markets where government incentives are limited or nonexistent. Continuously reducing production costs and increasing competition in the EV market are essential to making EVs more accessible.

The availability and accessibility of charging infrastructure remains a challenge. In many areas, the charging station network is not sufficiently developed to support mass EV adoption. The expansion of charging infrastructure, including fast charging in urban and rural areas, is crucial to provide consumers with confidence that they will be able to charge their vehicles conveniently.

Although EV range has improved significantly, there are still concerns about EVs' ability to make long journeys without frequent charging. In addition, charging times, although they are reducing, are still not as fast as refueling an internal combustion vehicle. Improvements in battery technology and fast-charging infrastructure are essential to address these concerns.

The production of lithium-ion batteries requires materials such as lithium, cobalt, and nickel, which have environmental and supply implications. Mining these materials can have negative environmental impacts, and the concentration of production in certain countries poses supply risks. The development of more sustainable alternatives, such as solid-state batteries and battery recycling, is crucial to mitigate these issues.

EV adoption also requires a change in consumer behavior. Many people are used to internal combustion vehicles and may be reluctant to switch to a new technology. Education and awareness of the benefits of EVs, as well as demonstrating their reliability and convenience, are essential to encourage adoption.

The future of electric vehicles

Despite the challenges, the future of electric vehicles (EVs) looks promising. With the continued advancement of technology, supportive policies, and growing consumer interest, EVs are well positioned to play a central role in the transformation of transportation.

Research and development in the field of batteries and electric motors continues to advance. Solid-state batteries, wireless charging, and the integration of renewables are just a few of the areas where significant progress is being made. These innovations have the potential to further improve EV performance, range and convenience.

Combining EVs with renewable energy sources is a key opportunity to further reduce transport emissions. EVs can be charged with electricity generated by solar, wind, and other renewable sources, creating a clean and sustainable energy cycle. In addition, EVs can play a role in stabilizing the electricity grid, acting as distributed energy storage and providing ancillary services to the grid.

Electric autonomous vehicles have the potential to transform urban transportation and reduce the need for private vehicle ownership. On-demand autonomous transport services could provide a convenient and sustainable alternative to the use of private cars, reducing congestion and emissions in cities.

Governments will continue to play a crucial role in promoting EVs. Policies that encourage EV adoption, charging infrastructure expansion, and renewable energy integration are essential to accelerate the transition. In addition, international cooperation in research and development, as well as in the standardization of technologies and regulations, can help overcome barriers and promote the global adoption of EVs.

Education and awareness of the benefits of EVs and sustainable transport in general are key to fostering behaviour change. Awareness campaigns can help consumers understand the advantages of EVs in terms of operating costs, environmental impact, and driving experience. Additionally, education about charging infrastructure and renewable energy options can increase confidence in EVs.

Conclusion

The history and evolution of electric vehicles is a narrative of innovation, challenges, and opportunities. From their earliest days in the 19th century to the current resurgence, EVs have come a long way. Today, they are at the heart of the transition to more sustainable and efficient transport.

Technological advances in batteries, electric motors, and energy management systems have been critical to the success of EVs. However, mass adoption of EVs requires overcoming several challenges, including upfront costs, charging infrastructure, and changing consumer behavior.

Despite these challenges, the future of EVs is bright. With continued technological innovation, public policy support, and growing consumer interest, EVs are well-positioned to transform transportation and contribute significantly to the reduction of greenhouse gas emissions.

This chapter has provided an overview of the history, technological advancements, and challenges of EVs. In the following chapters, we will explore in depth the innovations in public transport and shared mobility, the public policies that are facilitating the transition, and the case studies that demonstrate the positive impact of these technologies in different parts of the world. Through this exploration, we hope to provide a comprehensive and compelling understanding of how electric vehicles and sustainable transportation technologies can build a cleaner, more equitable future for all.

Chapter 2: Innovations in public transport and mobility

Public transport and shared mobility are essential components for the creation of a sustainable and efficient transport system. While electric vehicles are transforming individual transportation, public transportation and shared mobility solutions have the potential to significantly reduce reliance on private cars, decrease urban congestion, and improve air quality. This chapter explores innovations in public transport and shared mobility, analysing how these technologies are evolving and their impact on urban sustainability.

Electric public transport: Electric buses and trams

Electric public transport is experiencing a renaissance as cities look to reduce their carbon emissions and improve air quality. Electric buses and trams are two of the most promising solutions in this area.

Electric buses are rapidly replacing diesel buses in many cities around the world. These vehicles offer numerous advantages, such as reduced emissions, lower operating costs and quieter operation. One of the pioneering cities in the adoption of electric buses is Shenzhen, in China. In 2017, Shenzhen became the first city in the world to fully electrify its bus fleet, with more than 16,000 electric buses in operation. This change has significantly reduced CO_2 emissions and improved air quality in the city.

Battery technology for electric buses has advanced considerably, with faster charging times and longer range. Some companies are developing fast-charging systems at bus terminals, where vehicles can recharge their batteries in a matter of minutes during scheduled stops. In addition, inductive charging technology is being tested in several cities, allowing buses to be charged wirelessly while on the move or stopped at specific stops.

Electric trams, or light trams, are another form of electric public transport that is gaining popularity. These rapid transit systems are energy-efficient and can carry large numbers of passengers along fixed routes. Cities such as Melbourne, in Australia, and Vienna, in Austria, have extensive tram networks that form the backbone of their public transport systems.

Electric trams are not only more sustainable than internal combustion vehicles, but also contribute to urban revitalization. By eliminating the need for dedicated bus lanes and reducing car traffic, streetcars can help create more walkable and enjoyable streets for pedestrians and cyclists.

Shared mobility: Carsharing, bikesharing and ridesharing

Shared mobility solutions are revolutionizing the way people get around in cities. These options allow users to access vehicles when they need them, without the costs and responsibilities associated with owning a vehicle.

Carsharing, or carpooling, allows people to rent vehicles for short periods of time, usually by the hour or minute. This solution is ideal for those who need a car occasionally, but don't want the costs and maintenance associated with ownership. Companies such as Zipcar, Car2Go and, more recently, mobility platforms such as Turo, have popularized carsharing in many cities around the world. Carsharing can significantly reduce the number of vehicles on the streets, as a single carpool can replace several private cars. Not only does this decrease congestion and emissions, but it also frees up public space for other uses, such as parks and bike paths.

Bikesharing, or bicycle sharing, is another innovation that is transforming urban mobility. This system allows users to borrow bicycles from stations distributed throughout the city and return them at any other station. Cities such as Amsterdam, Copenhagen and Paris have led the implementation of bikesharing programs, promoting the use of bicycles as a healthy and sustainable alternative to the car. Electric bicycles, or e-bikes, are adding a new dimension to bikesharing. These motor-assisted bikes make it easier to get around on difficult terrain and long distances, making bike use accessible to a wider variety of people. In addition, the integration of smart technologies, such as GPS and mobile apps, has improved the ease of use and convenience of bikesharing.

Ridesharing, or ridesharing, allows users to share a car ride with others heading in the same direction. Platforms like Uber and Lyft have popularized ridesharing, providing a flexible and convenient alternative to using the private car. Ridesharing can reduce the number of vehicles on the roads, decrease congestion, and reduce per capita emissions. Ridesharing is also evolving with the integration of autonomous vehicles. Companies like Waymo and Cruise are developing fleets of autonomous vehicles for ridesharing services, which could further improve efficiency and reduce operating costs. In addition, the combination of ridesharing and electric vehicles can maximize environmental benefits, creating a more sustainable and efficient transportation system.

Autonomous vehicles and their integration

Autonomous vehicles, also known as driverless vehicles or self-driving cars, represent one of the most transformative innovations in the transportation sector. These vehicles use a combination of sensors, cameras, radars, and artificial intelligence algorithms to navigate and operate without human intervention.

Autonomous vehicles have the potential to significantly improve road safety, reduce congestion and decrease emissions. By eliminating human error, which is responsible for most road accidents, autonomous vehicles can make roads safer. In addition, autonomous vehicles can communicate with each other and with road infrastructure, optimizing traffic flow and reducing travel times.

The integration of autonomous vehicles with public transport can transform urban mobility. Autonomous vehicles can complement existing public transport systems, providing "last mile" solutions that connect users to train, metro and bus stations. This can improve the accessibility and convenience of public transportation, incentivizing more people to use it. For example, several cities are testing autonomous buses that operate on fixed routes or in specific areas. These buses can operate continuously and efficiently, without the need for breaks for drivers. In Helsinki, Finland, successful trials of autonomous buses have been conducted in residential areas and university campuses, demonstrating their feasibility as a complement to public transport.

Autonomous vehicles are also driving the development of on-demand mobility services. Companies like Waymo are testing fleets of autonomous vehicles that can be requested through mobile apps to provide on-demand rides. These services can reduce the need for private vehicle ownership and offer a convenient and sustainable alternative to car use.

Impact of shared mobility

Shared mobility offers numerous environmental and social benefits, but it also faces challenges that need to be addressed to maximize its positive impact.

Shared mobility can significantly reduce greenhouse gas emissions and air pollution. By reducing the number of vehicles on the roads, carsharing and ridesharing decrease congestion and improve transportation efficiency. In addition, bikesharing and the use of electric vehicles in shared mobility services contribute to a further reduction in emissions.

Shared mobility can also improve accessibility and equity in transportation. By providing affordable and flexible transportation options, shared mobility can benefit low-income communities and people who don't have access to a private car. In addition, by reducing the need for parking space, shared mobility can free up urban space for other uses, such as parks, public spaces, and housing.

Despite its benefits, shared mobility faces several challenges. One of the main challenges is competition with traditional public transport. In some cities, ridesharing has led to a decline in public transit use, which can undermine efforts to reduce congestion and emissions. It's important to find a balance between shared mobility and public transport to ensure that the two options complement each other.

Another challenge is regulation and safety. The rapid expansion of shared mobility services has in many cases outpaced the ability of regulators to monitor and ensure the security of these services. It is essential to develop regulatory frameworks that promote the safety, equity, and sustainability of shared mobility.

Policy Policies in Sustainable Technologies

Public policies play a crucial role in promoting sustainable transport technologies. Governments can implement a variety of measures to encourage the use of shared mobility and electric public transport solutions.

Subsidies and tax incentives can make sustainable technologies more affordable for consumers and businesses. For example, some governments offer subsidies for the purchase of electric vehicles and bicycles, as well as tax incentives for carsharing and bikesharing companies. These incentives can reduce the total cost of ownership and operation of sustainable technologies, increasing their adoption.

Regulations and standards are also critical to promoting sustainable transportation. Low emission zones, which restrict the access of high-emission vehicles to urban areas, can incentivize the use of electric vehicles and modes of shared transportation. In addition, regulations that set emission standards for vehicles and fleets can force transportation companies to adopt cleaner technologies.

Investments in infrastructure are essential to support the adoption of sustainable technologies. Governments can finance the construction of electric vehicle charging stations, bike lanes, and electric tram and bus networks. Adequate infrastructure is critical to ensuring that sustainable transport solutions are convenient and accessible to all users.

Public-private partnerships can accelerate the implementation of sustainable mobility solutions. Governments can work with technology and transportation companies to develop and test new solutions, share data and best practices, and fund pilot projects. These collaborations can foster innovation and the adoption of sustainable technologies.

To illustrate how innovations in public transport and shared mobility are being implemented and the impacts they are having, let's examine several case studies from different parts of the world.

Shenzhen is a prominent example of the electrification of public transport. In 2017, Shenzhen fully electrified its bus fleet, becoming the first city in the world to do so. The city now operates more than 16,000 electric buses, which has significantly reduced CO_2 emissions and improved air quality. The transition was made possible by a combination of government grants, investments in charging infrastructure, and collaboration with electric bus manufacturers.

Copenhagen is known for its cycling culture and bikesharing innovations. The city has implemented a bikesharing system that includes electric bicycles, which facilitates travel over difficult terrain and long distances. The integration of electric bicycles has increased the accessibility of bikesharing, attracting a greater diversity of users. In addition, Copenhagen has invested in an extensive network of cycle paths and promoted policies that encourage the use of bicycles as a primary mode of transport.

Helsinki has been a pioneer in the implementation of autonomous buses and on-demand mobility services. The city has successfully tested autonomous buses in residential areas and university campuses, demonstrating their viability as a complement to public transportation. In addition, Helsinki has launched an on-demand mobility service called "Whim", which integrates different modes of transport, including buses, bike-sharing, taxis, and ridesharing vehicles, into a single app. This comprehensive approach facilitates trip planning and improves the accessibility of sustainable transportation.

Bogota is a notable example of how a bus rapid transit (BRT) system can transform urban mobility. The TransMilenio system, launched in 2000, has significantly improved the efficiency of public transport in the city. TransMilenio uses dedicated bus lanes and elevated boarding stations, reducing travel times and improving punctuality. This system has eased congestion, reduced emissions and provided a viable alternative to the use of private cars.

Paris has been one of the leading cities in the implementation of bikesharing on a large scale. The Vélib' programme, launched in 2007, has been a resounding success, with thousands of bicycles available at stations throughout the city. Vélib' has promoted the use of bicycles as a means of daily transport, reducing dependence on cars and improving air quality. In addition, Paris has invested in cycling infrastructure, including protected bike lanes and secure parking, to support the growth of bikesharing.

Stockholm has implemented a low-emission zone in its city center, restricting access for high-emission vehicles and promoting the use of electric vehicles and shared mobility solutions. The city has seen a significant reduction in emissions and an improvement in air quality. In addition, Stockholm has promoted carsharing and ridesharing as alternatives to the use of private cars, which has contributed to reducing congestion and improving transport efficiency.

Future Perspectives: Innovations and Emerging Trends

The future of public transport and shared mobility is full of opportunities and challenges. As technologies continue to evolve and cities adopt more integrated and sustainable approaches, we can expect to see a number of innovations and trends emerging.

Hydrogen is emerging as a viable alternative to electric batteries for certain types of public transport, especially for heavy-duty vehicles such as buses and trucks. Hydrogen vehicles generate electricity through a chemical reaction between hydrogen and oxygen, emitting only water as a byproduct. Hydrogen offers a longer range and faster recharging times compared to electric batteries, making it an attractive option for long routes and intensive operations.

The integration of renewable energies into public transport and shared mobility is a growing trend. Electric vehicles can be charged with electricity generated by solar, wind, and other renewable sources, creating a clean and sustainable energy cycle. In addition, charging stations can be equipped with solar panels and energy storage systems, improving the resilience and sustainability of the transport system.

Autonomous vehicles have the potential to transform public transport and shared mobility. On-demand autonomous transportation services can provide a convenient and sustainable alternative to private car use, reducing congestion and emissions. In addition, autonomous buses and trams can improve the efficiency and punctuality of public transportation, providing a more reliable and accessible travel experience.

Mobility as a Service (MaaS) is an integrated approach that combines different modes of transportation on a single platform, allowing users to plan, book, and pay for rides using a single app. MaaS facilitates trip planning and improves the accessibility of sustainable transport, incentivising more people to use shared mobility and public transport solutions. This approach has the potential to transform the way people get around, making transportation more efficient, convenient, and sustainable.

Smart infrastructure, equipped with sensors and communication technologies, can improve the efficiency and sustainability of public transport and shared mobility. Smart transport networks can optimise traffic flow, improve the punctuality of public transport and provide real-time information to users. In addition, smart infrastructure can facilitate the integration of autonomous vehicles and wireless charging technologies, improving the convenience and accessibility of sustainable transportation.

Conclusion

Public transport and shared mobility are key elements in the transition to a more sustainable and efficient transport system. Innovations in electric buses and trams, as well as shared mobility solutions such as carsharing, bikesharing, and ridesharing, are transforming the way people get around in cities.

Public policies, infrastructure investments, and public-private partnerships are essential to support the adoption of these technologies and maximize their positive impact. Case studies from around the world show that with the right mix of policies, technology and political will, meaningful change is possible.

The future of public transport and shared mobility is full of opportunities. Technological innovations, such as autonomous vehicles, hydrogen as a fuel, and mobility as a service, have the potential to transform urban mobility and create a cleaner, more efficient, and equitable transportation system.

This chapter has explored innovations in public transport and shared mobility, providing an overview of their impact and potential. In the following chapters, we will continue to explore the public policies that are facilitating the transition and the case studies that demonstrate the positive impact of these technologies in different parts of the world. Through this exploration, we hope to provide a comprehensive and compelling understanding of how public transport and shared mobility can contribute to a more sustainable and equitable future for all.

Chapter 3: Policy Impact on Sustainable Technologies

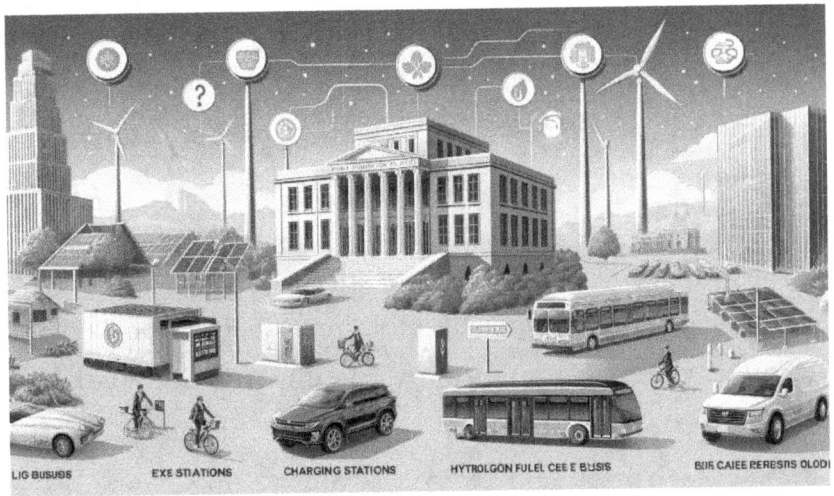

Public policies are an essential component in promoting sustainable transport technologies and facilitating the transition to a more efficient and greener transport system. Through a combination of economic incentives, regulations, infrastructure investments, and public-private partnerships, governments can create an enabling environment for the adoption of electric vehicles (EVs), electric public transportation, and shared mobility solutions. This chapter explores the various public policies implemented in different parts of the world, analyzing their impact on the adoption of sustainable technologies and providing examples of success.

Subsidies and economic incentives

The future of public transport and shared mobility is full of opportunities and challenges. As technologies continue to evolve and cities adopt more integrated and sustainable approaches, we can expect to see a number of innovations and trends emerging.

Hydrogen is emerging as a viable alternative to electric batteries for certain types of public transport, especially for heavy-duty vehicles such as buses and trucks. Hydrogen vehicles generate electricity through a chemical reaction between hydrogen and oxygen, emitting only water as a byproduct. Hydrogen offers a longer range and faster recharging times compared to electric batteries, making it an attractive option for long routes and intensive operations.

The integration of renewable energies into public transport and shared mobility is a growing trend. Electric vehicles can be charged with electricity generated by solar, wind, and other renewable sources, creating a clean and sustainable energy cycle. In addition, charging stations can be equipped with solar panels and energy storage systems, improving the resilience and sustainability of the transport system.

Autonomous vehicles have the potential to transform public transport and shared mobility. On-demand autonomous transportation services can provide a convenient and sustainable alternative to private car use, reducing congestion and emissions. In addition, autonomous buses and trams can improve the efficiency and punctuality of public transportation, providing a more reliable and accessible travel experience.

Mobility as a Service (MaaS) is an integrated approach that combines different modes of transportation on a single platform, allowing users to plan, book, and pay for rides using a single app. MaaS facilitates trip planning and improves the accessibility of sustainable transport, incentivising more people to use shared mobility and public transport solutions. This approach has the potential to transform the way people get around, making transportation more efficient, convenient, and sustainable.

Smart infrastructure, equipped with sensors and communication technologies, can improve the efficiency and sustainability of public transport and shared mobility. Smart transport networks can optimise traffic flow, improve the punctuality of public transport and provide real-time information to users. In addition, smart infrastructure can facilitate the integration of autonomous vehicles and wireless charging technologies, improving the convenience and accessibility of sustainable transportation.

Regulations and standards

Regulations and standards can establish standards and requirements that promote the use of sustainable technologies and discourage the use of polluting technologies. These policies play a crucial role in the transition to cleaner and more efficient mobility, ensuring that both manufacturers and consumers adopt more sustainable practices.

Low Emission Zones (LEZs) are areas where access by high-emission vehicles is restricted to reduce air pollution. These zones force drivers to use low-emission vehicles, such as electric vehicles (EVs) and hybrid vehicles, if they wish to enter these areas. London, for example, implemented an LEZ in 2019 that charges a daily fee to vehicles that do not meet emissions standards. This measure has led to a significant reduction in air pollution levels and has incentivised drivers to switch to cleaner vehicles. The implementation of LEZs not only improves air quality, but also encourages a shift towards more sustainable transport technologies in the most congested and pollution-affected urban areas.

Governments can set emission standards for new vehicles sold in their jurisdiction. These standards may require progressive reductions in emissions of CO_2 and other pollutants, incentivizing manufacturers to develop and market cleaner vehicles. The European Union, for example, has set ambitious emissions reduction targets for car manufacturers, with significant fines for those who fail to meet the targets. This strategy has led to greater innovation in vehicle design and the implementation of more advanced technologies to reduce emissions.

Regulations can also target commercial fleets, such as taxis and delivery vehicles. For example, some cities have implemented requirements for taxi and delivery vehicle fleets to be fully electric by a certain date. In Amsterdam, all taxis and delivery vehicles in the city centre are expected to be electric by 2025. These regulations not only reduce emissions from these fleets, but also serve as an example for other cities and sectors, showing the benefits of the transition to electric vehicles.

Governments can set energy efficiency standards for buildings, including requirements for the installation of EV charging stations. In California, the Green Building Code requires that new residential and commercial buildings be equipped with EV charging infrastructure, making it easier for residents and employees to adopt electric vehicles. This type of regulation ensures that charging infrastructure grows in parallel with the adoption of electric vehicles, removing one of the most significant barriers to EV expansion.

In addition to these specific regulations, public policies can include tax incentives and direct subsidies for the purchase of electric vehicles and bicycles. For example, some governments offer subsidies that significantly reduce the initial cost of these vehicles, making them more affordable for consumers. These incentives can reduce the total cost of ownership and operation of sustainable technologies, increasing their adoption and accelerating the transition to cleaner transportation.

Another important aspect is investment in public and private infrastructure to support these technologies. Governments can fund the construction of fast-charging stations for electric vehicles, as well as the expansion of bike lanes and electric tram and bus networks. Adequate infrastructure is critical to ensure that sustainable transport solutions are convenient and accessible to all users, promoting their widespread adoption.

Public-private partnerships can also accelerate the implementation of sustainable mobility solutions. Governments can work with technology and transportation companies to develop and test new solutions, share data and best practices, and fund pilot projects. These collaborations can foster innovation and the adoption of sustainable technologies, leveraging the resources and knowledge of both sectors to achieve common goals.

Examples of these approaches can be found in several cities around the world. Shenzhen, China, has managed to fully electrify its bus fleet thanks to a combination of government subsidies, investments in charging infrastructure, and collaboration with electric bus manufacturers. Copenhagen, Denmark, has implemented a bikesharing system that includes electric bicycles, making it easier to travel over difficult terrain and long distances, and has promoted policies that encourage the use of bicycles as the main means of transport. Helsinki, Finland, has been a pioneer in the implementation of autonomous buses and on-demand mobility services, integrating different modes of transport into a single application to improve the accessibility of sustainable transport.

Bogota, Colombia, has transformed urban mobility with its TransMilenio bus rapid transit (BRT) system, which uses dedicated bus lanes and elevated boarding stations to improve the efficiency of public transportation. Paris, France, has led the implementation of bikesharing with its Vélib' programme, which has reduced car dependency and improved air quality, supported by robust cycling infrastructure. Stockholm, Sweden, has implemented a low-emission zone and promoted carsharing and ridesharing, helping to reduce congestion and improve transport efficiency.

Regulations and standards, along with infrastructure investments and public-private partnerships, are essential to promote the use of sustainable technologies and discourage the use of polluting technologies. Through well-designed policies and the implementation of technological innovations, it is possible to move towards a future of cleaner, more efficient and accessible mobility for all.

Infrastructure investments

The right infrastructure is essential for the success of sustainable technologies. Governments can play a crucial role in financing and developing the infrastructure needed to support the adoption of electric vehicles (EVs), electric public transport, and shared mobility solutions.

A robust network of charging stations is critical to EV adoption. Governments can fund the construction of public and private charging stations, as well as fast-charging infrastructure on roads and urban areas. Norway, a global leader in EV adoption, has invested significantly in fast-charging stations, ensuring EV drivers can conveniently charge their vehicles. These investments not only increase consumer confidence in the viability of EVs, but also reduce one of the most significant barriers to mass adoption.

The electrification of public transport requires investments in infrastructure, such as charging stations for electric buses and tram lines. In China, the government has funded the construction of charging stations for electric buses in cities across the country, easing the transition to electric bus fleets. This infrastructure is crucial to ensure that electric buses can operate efficiently and continuously, reducing emissions and improving air quality in urban areas.

Investments in bike lanes and tram networks can promote the use of sustainable modes of transport. Cities such as Copenhagen and Amsterdam have invested in extensive networks of protected cycle paths, which has led to a significant increase in bicycle use. These investments not only improve cyclist safety, but also encourage a healthier lifestyle and reduce traffic congestion. In addition, investments in tram networks can improve the efficiency and accessibility of public transport, reducing reliance on the private car and offering a viable and sustainable alternative for daily commuting.

Infrastructure for shared mobility, such as bikesharing stations and carsharing parking areas, is also important. Governments can provide funding for the installation of bikesharing stations and work with carsharing companies to develop dedicated parking areas. In Paris, the government has supported the expansion of the Vélib' bikesharing program and worked with carsharing companies to facilitate car-sharing parking. These initiatives not only make shared mobility options more accessible, but also help reduce traffic congestion and emissions, promoting more efficient and sustainable urban mobility.

Investment in infrastructure is crucial to support the adoption of sustainable transport technologies. Governments have a critical role to play in financing and developing this infrastructure, ensuring that sustainable mobility solutions are convenient and accessible to all users. By creating an appropriate network of EV charging stations, electrifying public transport, building bike lanes and tram networks, and supporting shared mobility infrastructure, it is possible to promote a cleaner, more efficient and sustainable transport future.

Public-private partnerships

Public-private partnerships can accelerate the implementation of sustainable mobility solutions and foster innovation. These partnerships are crucial to developing the infrastructure and technologies needed for cleaner, more efficient mobility.

Public-private partnerships can be effective in the development of electric vehicle (EV) charging stations. Governments can work with energy companies and vehicle manufacturers to finance and build charging stations. Tesla, for example, has collaborated with local governments and energy companies to develop its network of superchargers around the world. Not only do these partnerships facilitate the expansion of charging infrastructure, but they also ensure that EVs have access to reliable and convenient charging points, which is essential to foster mass adoption of these vehicles.

Public-private partnerships can also facilitate pilot projects and technology testing. In Helsinki, Finland, the city has worked with technology and transportation companies to test autonomous buses in residential areas and university campuses. These tests have provided valuable data on the viability of autonomous buses and helped improve the technology. By conducting these pilot projects, cities can assess the performance and uptake of new technologies in a controlled environment before deploying them on a larger scale.

The development of smart infrastructure, equipped with sensors and communication technologies, can improve the efficiency and sustainability of transport. Governments can work with technology companies to develop and test smart infrastructure, such as smart transportation networks and traffic management systems. In Singapore, the government has collaborated with technology companies to develop a smart transport network that optimises traffic flow and improves the punctuality of public transport. This smart infrastructure enables more efficient traffic management, reduces congestion and improves the experience of public transport users.

Public-private partnerships can facilitate the financing of shared mobility projects. Governments can provide seed funding and work with shared mobility companies to develop carsharing and bikesharing programs. In Washington D.C., the government has worked with bikesharing companies to fund and expand the Capital Bikeshare program, which has been a huge success. These initiatives not only promote shared mobility, but also reduce the need for private vehicle ownership, decrease traffic congestion, and improve urban air quality.

Public-private partnerships are essential to accelerate the implementation of sustainable mobility solutions. By jointly developing EV charging stations, conducting pilot projects and technology tests, creating smart infrastructure and financing shared mobility projects, it is possible to foster innovation and promote a cleaner and more efficient transport future. These alliances leverage the resources and knowledge of both sectors to achieve common goals, benefiting both citizens and the environment.

Impact of public policies on sustainable technologies

To illustrate how public policies can facilitate the adoption of sustainable technologies, let's examine several case studies from different parts of the world.

Norway is a world leader in the adoption of electric vehicles (EVs), thanks to a number of government incentives. EV buyers in Norway are exempt from purchase tax and VAT, which can reduce the price of an EV by up to 25%. In addition, EVs are exempt from tolls, parking fees, and highway tolls, making them cheaper to use. These policies have led to more than half of new cars sold in Norway being electric. This strategy has created a favorable environment for the mass adoption of EVs, transforming the country's automotive market.

China has made great strides in the electrification of public transport, especially electric buses. The Chinese government has provided generous subsidies to bus companies to encourage the transition to electric vehicles. In addition, Chinese cities have implemented regulations requiring the electrification of bus fleets. In Shenzhen, these subsidies and regulations have led to the complete electrification of the city's bus fleet. This initiative has not only reduced CO_2 emissions, but has also significantly improved air quality in urban areas.

London implemented a Low Emission Zone (LEZ) in 2019, which charges a daily fee to vehicles that do not meet emissions standards. This measure has led to a significant reduction in air pollution levels and has incentivised drivers to switch to cleaner vehicles. In addition, the London government has invested in EV charging infrastructure and the electrification of public transport. These investments have created a healthier urban environment and encouraged the adoption of cleaner transport technologies.

California is a leading example of how regulations and subsidies can promote the adoption of sustainable technologies. The state has set strict energy efficiency standards for buildings, including requirements for the installation of EV charging stations. In addition, California offers subsidies and tax credits for the purchase of EVs and the installation of charging infrastructure. These policies have led to a significant increase in EV adoption and charging infrastructure development in the state. Energy efficiency regulations also ensure that new buildings are sustainable and future-proof for electric mobility.

Copenhagen has invested significantly in cycling infrastructure, including an extensive network of protected cycle paths. The city has also implemented a successful bikesharing program that includes electric bicycles. These investments have led to a significant increase in the use of bicycles as a means of transportation, reducing dependence on the car and improving air quality. Safe and accessible cycling infrastructure has made Copenhagen a role model for other cities looking to promote sustainable mobility and reduce urban emissions.

These examples demonstrate how public policies can play a crucial role in promoting sustainable technologies. From financial incentives to strict regulations and investment in infrastructure, governments have a variety of tools at their disposal to encourage the adoption of clean technologies and improve the sustainability of transport systems. By implementing these policies, cities and countries can move towards a cleaner, more efficient, and sustainable future.

Challenges and lessons learned

Despite the success of many public policies, the promotion of sustainable technologies faces several challenges that must be addressed to ensure their adoption and effectiveness in the long term.

The upfront cost of sustainable technologies can be a significant barrier to adoption. Electric vehicles (EVs), for example, typically have a higher starting price compared to internal combustion vehicles. Governments must find ways to finance subsidies and infrastructure developments without compromising other essential public services. Public-private partnerships can be an effective solution, where private companies invest in the necessary infrastructure in exchange for tax incentives or subsidies. In addition, innovative financing mechanisms, such as green bonds, can provide the necessary funds to support these developments. These bonds allow investors to finance sustainability projects, providing a flow of capital that can be directed towards the construction of EV charging stations, the electrification of public transport and other similar initiatives.

Public acceptance and behaviour change are essential for the success of sustainable mobility policies. While technologies may be available and accessible, their adoption is highly dependent on people's willingness to change their habits and adopt new practices. Awareness and education campaigns can play a crucial role in this regard, helping consumers understand the benefits of sustainable technologies, both in terms of long-term savings and environmental impact. In addition, it is vital that policies are inclusive and equitable, ensuring that all segments of the population, including the most disadvantaged, can benefit from these technologies. This may involve targeted grants, accessible financing programs, and consideration of the specific needs of diverse communities.

The development and maintenance of infrastructure for sustainable technologies requires significant investments and long-term planning. It is not enough to install EV charging stations or build bike lanes; These infrastructures must be maintained and updated regularly to ensure their functionality and resilience. Governments must ensure that infrastructure is adequate and resilient, adapting to changing needs and withstanding intensive use. Integrated urban planning and coordination between different levels of government are crucial to the success of these initiatives. This implies a coherent and collaborative vision that considers urban growth, sustainability and infrastructure connectivity.

Regulations and standardization are essential to ensure the interoperability and security of sustainable technologies. Without clear standards, there can be confusion and difficulties for consumers, as well as technical challenges for suppliers. Governments should work closely with industry and international bodies to develop and adopt standards and regulations that promote the adoption of sustainable technologies. For example, standardizing EV charging stations is crucial to make them easier for consumers to use, ensuring that any electric vehicle can be charged at any available station. In addition, regulations must ensure that these technologies are safe and efficient, protecting both users and the environment.

While the challenges are significant, so are the opportunities to move towards a more sustainable mobility future. With a strategic and collaborative approach, it is possible to overcome barriers and encourage the adoption of sustainable technologies, ensuring environmental, economic and social benefits for all.

Innovations and trends in public policies

The future of public policies for sustainable mobility is full of opportunities and challenges. As technologies continue to evolve and cities adopt more integrated and sustainable approaches, we can expect to see a number of innovations and emerging trends that will transform the way we get around.

Mobility as a Service (MaaS) is an integrated approach that combines different modes of transportation on a single platform, allowing users to plan, book, and pay for rides using a single app. Governments can support the development of MaaS by working with technology and transportation companies to develop integrated platforms and provide incentives for their use. MaaS has the potential to transform the way people get around, making transportation more efficient, convenient, and sustainable. Imagine a future where you can use a single app to rent a bike, book a carpool, and pay for a train ticket, all in one transaction. This level of integration can reduce reliance on the private car, ease traffic congestion, and decrease emissions.

Governments can develop policies that promote the integration of renewable energy in transport. This can include incentives for the use of solar and wind-generated electricity to charge electric vehicles (EVs), as well as the installation of solar panels at charging stations and bus terminals. The integration of renewable energy can create a clean and sustainable energy cycle, further reducing transport emissions. For example, EV charging stations equipped with solar panels would not only provide clean energy for vehicles, but could also store energy in batteries for use during peak demand, improving the energy resilience of the grid.

The development of smart infrastructure, equipped with sensors and communication technologies, can improve the efficiency and sustainability of transport. Governments can support the development of smart transport networks, traffic management systems and smart charging stations. These technologies can optimize traffic flow, improve the punctuality of public transportation, and provide real-time information to users. Imagine a public transportation system where buses and trains coordinate seamlessly with traffic conditions and passenger schedules, minimizing wait times and maximizing system efficiency.

Governments can explore innovative financing mechanisms, such as green bonds and public-private partnerships, to finance the development of sustainable technologies and necessary infrastructure. Green bonds, which are used to finance environmental and sustainable projects, can provide a long-term source of financing for sustainable mobility initiatives. This type of financing allows cities to invest in large-scale projects, such as the expansion of electric public transport networks or the construction of protected bike lanes, without relying exclusively on the public budget.

Circular economy policies can promote the reuse and recycling of materials in the transport sector. Governments can develop regulations that require the recovery and recycling of EV batteries, as well as the reuse of materials in vehicle manufacturing. The circular economy can reduce the demand for natural resources and minimize the environmental impacts of vehicle production and disposal. For example, EV batteries that have reached the end of their useful life in cars can be recycled or reused in energy storage applications, extending their lifespan and reducing the need to extract new materials.

The future of public policies for sustainable mobility promises to be dynamic and transformative. MaaS implementation, renewable energy integration, smart infrastructure development, innovative financing, and circular economy regulations are just a few of the strategies that can help create more efficient, sustainable, and equitable transportation systems. By adopting these policies and fostering collaboration between the public and private sectors, governments can lead the way towards a cleaner and more efficient mobility future.

Conclusion

Public policies are an essential component in promoting sustainable transport technologies and facilitating the transition to a more efficient and greener transport system. Through a combination of subsidies, tax incentives, regulations, infrastructure investments, and public-private partnerships, governments can create an enabling environment for the adoption of EVs, electric public transport, and shared mobility solutions.

Case studies from different parts of the world show that, with the right mix of policies, technology and political will, meaningful change is possible. However, the promotion of sustainable technologies faces several challenges, such as initial cost, public acceptance, and infrastructure development. As technologies continue to evolve and cities adopt more integrated and sustainable approaches, we can expect to see a number of innovations and emerging trends in public policy.

This chapter has explored public policies and their impact on the adoption of sustainable technologies, providing examples of success and lessons learned. In the following chapters, we will continue to explore the case studies that demonstrate the positive impact of these technologies in different parts of the world. Through this exploration, we hope to provide a comprehensive and compelling understanding of how public policies can contribute to a more sustainable and equitable future for all.

Chapter 4: Case Studies and Future Perspectives

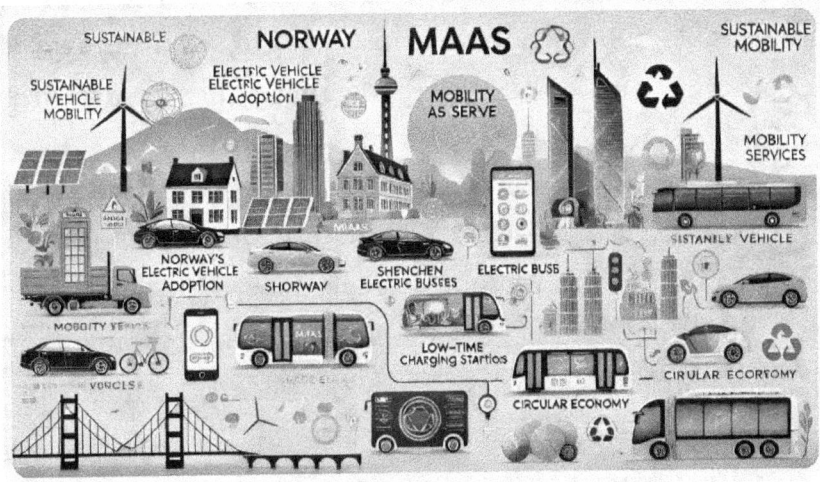

The impact of public policies, technological innovations and sustainable mobility solutions can be clearly seen in various case studies from around the world. These examples highlight successful strategies, challenges overcome, and lessons learned in the transition to more sustainable transportation. In addition, the future outlook offers insight into emerging trends and opportunities to continue advancing sustainable mobility. This chapter presents a series of case studies from different cities and regions, followed by an exploration of the future prospects for sustainable mobility.

Case Studies: Innovations and Successes in Sustainable Mobility

The future of public policies for sustainable mobility is full of opportunities and challenges. As technologies continue to evolve and cities adopt more integrated and sustainable approaches, we can expect to see a number of innovations and emerging trends that will transform the way we get around.

Norway is a world leader in the adoption of electric vehicles (EVs), with more than half of new cars sold in the country being electric. This success is due to a combination of favorable public policies, economic incentives, and a well-developed charging infrastructure. EV buyers in Norway are exempt from purchase tax and VAT, which can reduce the price of an EV by up to 25%. In addition, EVs are exempt from tolls, parking fees, and highway tolls, making them cheaper to use. These policies have led to a mass adoption of EVs, significantly reducing CO_2 emissions from transport and improving air quality in Norwegian cities. The charging infrastructure has expanded rapidly, with fast-charging stations available across the country, making everyday EVs easier to use. Norway has shown that with strong government support and adequate infrastructure, it is possible to achieve a rapid and successful transition to electric mobility. However, it has also faced challenges, such as the need to maintain and expand charging infrastructure and compensate for lost revenue from fossil fuel taxes. Norway's experience highlights the importance of long-term planning and adapting policies as the market changes.

Shenzhen, China, is a prominent example of the electrification of public transport. In 2017, Shenzhen fully electrified its bus fleet, becoming the first city in the world to do so. This achievement was made possible by generous government subsidies, investments in charging infrastructure and collaboration with electric bus manufacturers. The Chinese government has provided financial support and regulations mandating the electrification of bus fleets. The electrification of Shenzhen's bus fleet has significantly reduced CO_2 emissions and improved air quality in the city. Electric buses also offer lower operating costs and quieter operation compared to diesel buses. This shift has shown that the electrification of public transport is a viable and beneficial solution for large cities. However, the significant upfront investment required for the purchase of electric buses and the installation of charging infrastructure, as well as the need for training programs for drivers and technicians, have been significant challenges. Shenzhen's experience underscores the importance of government subsidies and collaboration with industry to overcome initial barriers and achieve a successful transition.

London, United Kingdom, has implemented several policies to reduce air pollution and promote sustainable mobility, including the Low Emission Zone (LEZ) and the Ultra Low Emission Zone (ULEZ). These zones charge a daily fee to vehicles that do not meet emissions standards, incentivizing drivers to switch to cleaner vehicles. In addition, the London government has invested in EV charging infrastructure and the electrification of public transport. LEZs and ULEZs have led to a significant reduction in air pollution levels and incentivised the adoption of cleaner vehicles. Investments in charging infrastructure and the electrification of public transport have improved the accessibility and efficiency of sustainable transport in London. These policies have helped transform urban mobility and improve the quality of life for residents. However, the implementation of LEZs and ULEZs has faced resistance from some drivers and industry sectors, and the expansion of charging infrastructure requires continued investment and careful planning. London has learned that communication and collaboration with all stakeholders are crucial to the success of these policies.

California, United States, has set strict energy efficiency standards for buildings and offers subsidies and tax credits for the purchase of EVs and the installation of charging infrastructure. The state has also implemented regulations mandating a progressive reduction in CO_2 emissions from new vehicles sold in the state. These policies have led to a significant increase in EV adoption and charging infrastructure development in California. Energy efficiency standards have promoted the construction of more sustainable buildings and the integration of EV charging stations. Emissions regulations have incentivized automakers to develop cleaner, more efficient vehicles. However, one of the challenges is the high upfront cost of sustainable technologies and the need for ongoing financing for subsidies and infrastructure. In addition, EV adoption varies between different regions and communities within the state. California has learned that it is important to provide incentives and support tailored to the needs of different demographic and geographic groups.

Copenhagen, Denmark, has invested significantly in cycling infrastructure, including an extensive network of protected cycle paths. The city has also implemented a successful bikesharing program that includes electric bicycles. These investments have been part of a broader strategy to promote the use of bicycles as the main means of transport. Investments in cycling infrastructure and bikesharing have led to a significant increase in the use of bicycles as a means of transport in Copenhagen. This has reduced dependence on the car, decreased emissions and improved air quality. The city has seen additional benefits in terms of public health and well-being, as more people opt for an active mode of transportation. However, one of the challenges has been the integration of bicycles with other modes of transport and the need to continuously maintain and expand cycling infrastructure. Copenhagen has learned that integrated urban planning and community engagement are crucial to the success of cycling mobility policies.

Helsinki, Finland, has been a pioneer in the implementation of autonomous buses and on-demand mobility services. The city has successfully tested autonomous buses in residential areas and university campuses, demonstrating their viability as a complement to public transportation. In addition, Helsinki has launched an on-demand mobility service called "Whim", which integrates different modes of transport into a single app. Testing of autonomous buses has provided valuable data on the viability of this technology and improved the efficiency of public transport. The on-demand mobility service has made trip planning easier and improved the accessibility of sustainable transport. These advances have demonstrated the potential of technology to transform urban mobility. However, one of the challenges has been the integration of emerging technologies with existing transport systems and the need to develop appropriate regulatory frameworks. Helsinki has learned that collaboration with technology and transport companies, as well as public participation, are essential to the success of these innovations.

Future Outlook: Emerging Trends and Opportunities

The future of sustainable mobility is full of opportunities and challenges. As technologies continue to evolve and cities adopt more integrated and sustainable approaches, we can expect to see a number of emerging trends that will transform urban mobility.

Mobility as a Service (MaaS) is an integrated approach that combines different modes of transportation on a single platform, allowing users to plan, book, and pay for rides using a single app. MaaS facilitates trip planning and improves the accessibility of sustainable transport, incentivising more people to use shared mobility and public transport solutions. MaaS has the potential to transform the way people get around, making transportation more efficient, convenient, and sustainable. By integrating different modes of transportation, MaaS can reduce reliance on the private car and promote the use of sustainable mobility solutions. However, one of the challenges is the integration of different transport systems and the need to develop interoperable technological platforms. In addition, it is important to ensure that MaaS is accessible and affordable for all users, including low-income communities and rural areas.

The integration of renewable energies in transport is a growing trend. Electric vehicles (EVs) can be charged with electricity generated by solar, wind, and other renewable sources, creating a clean and sustainable energy cycle. In addition, charging stations can be equipped with solar panels and energy storage systems. The integration of renewable energies can significantly reduce transport emissions and improve the sustainability of the energy system. Renewable energy charging stations can improve the resilience of charging infrastructure and reduce dependence on the power grid. However, one of the challenges is the initial cost of installing renewable energy systems and the need for financing. In addition, it is important to develop policies and regulations that promote the integration of renewable energies in transport and ensure the interoperability of charging systems.

The development of smart infrastructure, equipped with sensors and communication technologies, can improve the efficiency and sustainability of transport. Smart transport networks can optimise traffic flow, improve the punctuality of public transport and provide real-time information to users. Smart infrastructure can significantly improve transportation efficiency, reduce congestion and emissions, and facilitate the integration of autonomous vehicles and wireless charging technologies, improving the convenience and accessibility of sustainable transportation. However, one of the challenges is the cost of developing and implementing smart infrastructure and the need for collaboration between different levels of government and the private sector. In addition, it is crucial to ensure the security and privacy of the data collected by the smart infrastructure.

Autonomous vehicles and on-demand mobility have the potential to transform urban mobility. On-demand autonomous transportation services can provide a convenient and sustainable alternative to private car use, reducing congestion and emissions. Autonomous vehicles can improve road safety, reduce operating costs, and increase transportation efficiency. On-demand mobility can offer flexible and accessible solutions, complementing public transport and reducing the need for private vehicle ownership. However, one of the challenges is the development and implementation of appropriate regulatory frameworks for autonomous vehicles. In addition, it is important to ensure public acceptance and trust in autonomous technology. Integrating autonomous vehicles with existing transport systems also presents technical and logistical challenges.

Circular economy policies can promote the reuse and recycling of materials in the transport sector. Governments can develop regulations that require the recovery and recycling of EV batteries, as well as the reuse of materials in vehicle manufacturing. The circular economy can reduce the demand for natural resources and minimize the environmental impacts of vehicle production and disposal.

In addition, it can create new economic opportunities in recycling and reusing materials. However, one of the challenges is the development of efficient technologies and processes for the recovery and recycling of materials. In addition, it is important to establish regulations and standards that promote the circular economy and ensure cooperation between different industrial sectors.

The future of sustainable mobility promises to be dynamic and transformative. Implementing MaaS, integrating renewable energy, developing smart infrastructure, adopting autonomous vehicles, and promoting the circular economy are just a few of the strategies that can help create more efficient, sustainable, and equitable transportation systems. By adopting these policies and fostering collaboration between the public and private sectors, governments can lead the way towards a cleaner and more efficient mobility future.

Conclusion

The case studies presented in this chapter highlight how diverse cities and regions have implemented public policies and adopted sustainable technologies to transform urban mobility. These examples demonstrate that, with the right mix of policies, technology and political will, it is possible to achieve meaningful change towards a more sustainable and efficient transport system.

As we move into the future, emerging trends and opportunities in sustainable mobility offer a promising path to continue improving the efficiency and sustainability of transport. Mobility as a service, the integration of renewable energies, the development of smart infrastructure, autonomous vehicles and the circular economy are just some of the trends that are shaping the future of mobility.

To capitalize on these opportunities, it is essential that governments, businesses, and communities work together to develop and adopt innovative solutions. Collaboration, long-term planning and a commitment to sustainability will be crucial to ensure that future generations can enjoy a clean, efficient and equitable transport system.

This chapter has explored case studies and future prospects in sustainable mobility, providing a comprehensive view of successful strategies and emerging opportunities. In the following chapters, we will continue to delve into the technologies and policies that are shaping the future of mobility, with the aim of providing a comprehensive and compelling understanding of how we can build a more sustainable and equitable future for all.

Chapter 5: Technological Innovation in Sustainable Mobility

Technological innovation is a crucial driver for the transformation of mobility towards more sustainable models. Over the past decade, we've seen impressive progress in areas such as electric vehicles, charging infrastructure, telematics systems, and alternative fuels. This chapter will delve into these innovations, exploring how they are changing the transport landscape and what their potential long-term impacts are.

New generation electric vehicles

Electric vehicles (EVs) have come a long way since their first versions. Today, innovations continue to improve its performance, autonomy and accessibility, marking a before and after in the automotive industry.

Batteries are at the heart of EVs, and improvements in their technology have been critical to their mass adoption. Lithium-ion batteries have dominated the market due to their high energy density and efficiency. However, research continues to move forward to overcome its limitations, such as costs and degradation over time. One of the most promising areas is the development of solid-state batteries, which use a solid electrolyte instead of a liquid one. These batteries promise higher energy density, faster charging times, and increased safety. Companies such as Toyota and QuantumScape are leading research in this area, and the first vehicles with these batteries are expected to hit the market in the coming years. In addition, efficient battery recycling is essential for the long-term sustainability of EVs. Advanced recycling methods are being developed to recover valuable materials such as lithium, cobalt and nickel, reducing the need for re-extraction and minimizing environmental impact.

In addition to batteries, electric motors and propulsion systems are also evolving. Permanent magnet motors offer high efficiency and performance thanks to the use of rare earth magnets. Improvements in the manufacture and design of these engines are leading to lighter and more efficient vehicles. To reduce reliance on rare materials, some manufacturers are developing magnet-free motors that use alternative technologies such as the switched reluctance motor. These engines can be more economical and sustainable in the long term. Power electronics are improving the efficiency of electric motors. More sophisticated motor controllers allow for more precise energy management, optimizing the performance and range of vehicles.

EV design and manufacturing are also evolving to be more sustainable. Manufacturers are adopting recycled materials and lightweight composites to reduce the weight of vehicles and improve their efficiency. For example, BMW uses recycled carbon fiber in its i3 model. 3D printing and other additive manufacturing techniques are enabling the production of lighter and more complex components, reducing waste and improving manufacturing efficiency. In addition, the concept of circular economy is being integrated into EV manufacturing, with strategies to reuse components and materials at the end of the vehicle's life. Not only does this reduce environmental impact, but it can also offer economic benefits.

Advances in battery technology, powertrains, and manufacturing practices are driving EV development and adoption. These innovations not only improve vehicle performance and efficiency, but also contribute to long-term environmental and economic sustainability. With continued support from research and development, as well as the adoption of enabling policies, EVs are well positioned to play a crucial role in the transition to a cleaner and more efficient mobility future.

Wireless and fast charging

Charging infrastructure is crucial to the success of electric vehicles (EVs), and innovations in this area are improving convenience and accessibility, facilitating the mass adoption of these vehicles.

Fast-charging stations allow EVs to recharge their batteries in minutes instead of hours, which is essential for widespread adoption. The new fast-charging stations can deliver up to 350 kW of power, allowing an EV's battery to be charged to 80% in less than 20 minutes. Networks such as Ionity in Europe and Electrify America in the United States are deploying these stations on a large scale, making fast charging available in more places and improving the user experience. To handle the high charging power without overheating the battery, advanced cooling technologies are being developed. These systems allow batteries to be kept at an optimal temperature during fast charging, prolonging their lifespan and ensuring safe performance.

Wireless charging promises to eliminate the need for physical connectors, improving convenience and making it easier to charge on the go. Inductive charging stations use electromagnetic fields to transfer energy between a charging pad on the ground and a receiving coil in the vehicle. This technology is ideal for urban applications and parking lots, where convenience is key. In addition, dynamic loads allow EVs to be charged while on the move, using coils built into the roads. Although still in the experimental phase, this technology could revolutionize long-range transportation, eliminating the need for long stops to recharge and increasing the efficiency of the transportation system.

Integrating renewables into charging infrastructure is crucial to maximizing the environmental benefits of EVs. Charging stations equipped with solar panels can generate their own electricity, reducing reliance on the grid and decreasing carbon emissions. Companies such as Envision Solar are developing mobile and autonomous solar charging stations, which can be deployed in various locations without the need for additional electrical infrastructure. To manage the variability of renewable energies, energy storage systems are being integrated into charging stations. Stationary batteries can store excess solar or wind energy during peak production and release it when demand is high, ensuring a constant and reliable power supply.

Innovations in charging infrastructure are playing a crucial role in facilitating EV adoption. Fast charging stations, wireless charging, and renewable energy integration not only improve the convenience and accessibility of EVs, but also contribute to a more sustainable and efficient future. With the continuous development and deployment of these technologies, the transition to wider and more accessible electric mobility is getting closer and closer to becoming a reality.

Telematics and connected vehicles

Telematics and connectivity are transforming vehicle management and operation, improving efficiency, safety, and convenience on multiple levels.

Telematics allows companies to manage their fleets more efficiently, reducing costs and improving performance. Telematics systems can track the location, vehicle status, and driver behavior in real-time. This allows companies to optimize routes, reduce fuel consumption, and improve safety. In addition, telematics facilitates predictive maintenance, identifying potential problems before they become costly failures. Built-in sensors can monitor wear on key components and alert fleet managers when maintenance is needed. This monitoring and preventative maintenance capability not only saves money, but also ensures that vehicles are always in optimal operating condition.

Autonomous vehicles promise to improve road safety, reduce congestion and increase transport efficiency. These vehicles use a combination of cameras, radars, lidars, and ultrasonic sensors to perceive their surroundings. These sensors enable precise and safe driving, even in adverse conditions. Artificial intelligence algorithms analyze sensor data to make real-time decisions. These systems are designed to continuously improve through machine learning, adapting to new situations and improving their performance over time. In addition, vehicle-to-everything (V2X) communication networks allow autonomous vehicles to communicate with each other and with road infrastructure. This facilitates traffic coordination, intersection management, and rapid response to emergencies, creating a safer and more efficient transportation environment.

Connectivity is also transforming the user experience through advanced infotainment systems. Modern infotainment systems allow full integration with smartphones, offering seamless navigation, entertainment and connectivity. Apple's CarPlay and Google's Android Auto are popular examples of these integrations, allowing users to access their favorite apps and services directly from the vehicle's console. In addition, AI-based virtual assistants, such as Amazon's Alexa and Google Assistant, are being integrated into infotainment systems. These assistants allow drivers to control vehicle functions, search for information and access services with voice commands, improving convenience and safety by reducing the need to physically interact with devices while driving.

Connected vehicles can receive over-the-air (OTA) software updates, improving their features and correcting problems without the need for workshop visits. Tesla has led the way in this area, proving that OTA updates can keep vehicles up to date with the latest innovations. Not only do these updates improve the vehicle's functionality and performance, but they can also introduce new features and safety enhancements, providing vehicle owners with a continuously improved user experience.

Telematics and connectivity are revolutionizing the automotive industry. From efficient fleet management and predictive maintenance to autonomous driving and advanced infotainment systems, these technologies are improving every aspect of transportation. With the continuous integration of these innovations, the future of mobility promises to be safer, more efficient and more convenient for everyone.

Emission reduction technologies

In addition to electric vehicles, there are other technologies that are helping to reduce transport emissions, offering alternatives to improve efficiency and reduce environmental impact.

Alternative fuels offer an option to reduce emissions from internal combustion vehicles. Compressed natural gas (CNG) produces fewer CO_2 emissions and pollutants compared to gasoline and diesel. It is mainly used in commercial and public transport fleets, where its lower cost and lower emissions can have a significant impact. Biofuels, such as ethanol and biodiesel, are produced from renewable feedstocks. These fuels can significantly reduce carbon emissions when produced sustainably, leveraging agricultural waste and other renewable sources. Hydrogen is a promising option for heavy-duty and long-range transportation. Hydrogen vehicles emit only water vapour and can be recharged quickly. Hydrogen fuel cells are being developed by companies such as Toyota, Honda and Hyundai, and are expected to play an important role in the decarbonisation of transport.

Plug-in hybrid vehicles (PHEVs) combine an internal combustion engine with an electric motor and a rechargeable battery, allowing for greater range and flexibility. PHEVs offer the possibility of driving short distances in all-electric mode, reducing emissions in urban environments. For longer distances, the internal combustion engine provides greater range, offering the best of both worlds. Manufacturers are developing PHEVs with higher-capacity batteries and more powerful electric motors, which improves efficiency and pure electric drivability. This allows users to enjoy greener driving without worrying about limited range.

Although electric vehicles are gaining traction, internal combustion engines will continue to be an important part of the automotive landscape for the foreseeable future. Therefore, technologies that reduce emissions from these engines are crucial. Modern exhaust systems, such as particulate filters and advanced catalytic converters, can significantly reduce pollutant emissions. These systems are continually being improved to meet stricter regulations, helping to minimize the environmental impact of combustion engines. Direct fuel injection and high-pressure injection are improving engine efficiency and reducing emissions. These technologies allow for fuller, cleaner combustion, increasing efficiency and reducing fuel consumption.

Mild hybridization uses a small electric motor to support the internal combustion engine. This reduces fuel consumption and emissions, especially during starting and accelerating. Mild hybrid systems are being adopted by many manufacturers as an intermediate solution to improve fuel efficiency and reduce emissions without the cost and complexity of a full hybrid system.

In addition to electric vehicles, a variety of alternative technologies and fuels are playing a crucial role in reducing transport emissions. From alternative fuels such as CNG and biofuels, to advanced technologies in combustion engines and hybrid systems, these innovations are helping to create a cleaner and more sustainable mobility future. With the continued development and adoption of these technologies, we can expect to see significant improvements in transportation efficiency and sustainability in the coming years.

Conclusion

Technological innovation is playing a key role in transforming mobility towards more sustainable models. From advances in EV batteries and powertrains, to improvements in charging infrastructure, telematics, and alternative fuels, these technologies are making transportation cleaner, more efficient, and more accessible.

Investment in research and development is crucial to keep pace with innovation and overcome technological barriers. Lessons learned in this area show that continued investment is essential to keep moving forward. Collaboration between governments, businesses and universities is essential to accelerate the development and implementation of new technologies. This cooperation allows for the sharing of knowledge, resources and experiences, resulting in faster and more effective progress. Policies and strategies must be adaptive, allowing for adjustments as technologies and market needs evolve.

Flexibility in policy-making ensures that innovations can be quickly adopted and applied in the right context.

The future prospects in the field of sustainable mobility are promising. Solid-state batteries and other emerging technologies promise to improve the range, safety, and sustainability of electric vehicles (EVs). These batteries have the potential to offer higher energy densities, faster charging times, and increased safety, which could revolutionize the EV market. Wireless and dynamic charging also promises to revolutionize the way EVs are charged, improving convenience and accessibility. With these technologies, vehicles could be charged without the need for physical connectors, and even while on the move, eliminating the need for long stops to recharge.

Autonomous and connected vehicles are transforming traffic management and user experience, making transportation safer and more efficient. Autonomous driving, supported by communication networks and advanced sensors, can reduce traffic accidents and optimize the flow of vehicles on the roads. In addition, connectivity allows for better fleet management and a more personalized and comfortable user experience. Adopting circular economy principles in the manufacturing and recycling of components can reduce the environmental impact of transportation and create new economic opportunities. This involves designing vehicles and components that can be easily disassembled and reused, minimizing waste and making the most of resources.

The future of sustainable mobility is promising, and technological innovation will be key to achieving it. With the right approach, we can build a transportation system that not only meets our mobility needs, but also protects our planet for future generations. The combination of continuous investments in R+D, collaboration between different sectors and the adoption of flexible and adaptive policies will allow us to meet the challenges and seize the opportunities that the future of sustainable mobility holds for us.

Chapter 6: Sustainable Mobility in Smart Cities

Smart cities are at the heart of the sustainable mobility revolution. By integrating advanced technologies and innovative urban planning approaches, smart cities seek to improve the quality of life of their inhabitants, reduce environmental impact, and create more efficient and sustainable transportation systems. This chapter will explore how smart cities are transforming urban mobility, with a focus on key technologies, integration models, and successful examples from around the world.

Definition and characteristics of a smart city

A smart city uses information and communication technologies (ICT) to improve the efficiency of urban services, reduce costs and resources, and increase the quality of life of its citizens. Mobility is a core component of smart cities, and its integration is crucial to achieving sustainability and efficiency goals.

The essential elements of a smart city include a digital infrastructure, IoT sensors and devices, Big Data and analytics, and citizen participation. Digital infrastructure consists of advanced communication networks that enable connectivity between devices, systems, and people. This is critical to the operation of a smart city, as it ensures that information is transmitted quickly and efficiently. Internet of Things (IoT) sensors and devices collect and analyze data in real-time, providing valuable insights into the state of urban services and citizen behavior. This data is analysed using Big Data and analytics techniques to make informed decisions and optimise urban services. Citizen participation is achieved by involving citizens in planning and decision-making through digital platforms and feedback mechanisms, ensuring that the solutions implemented respond to the needs and expectations of the community.

The technologies used in smart cities are varied and advanced, including 5G communication networks, open data platforms, artificial intelligence (AI), and blockchain.

5G communication networks offer greater speed and capacity, enabling real-time connectivity for autonomous vehicles and traffic management systems. This is crucial for the implementation of advanced mobility solutions that improve the efficiency of urban transport. Open data platforms enable access to urban data for developers and citizens, fostering innovation and transparency. By making data publicly available, it facilitates the development of applications and services that can improve life in the city. Artificial intelligence (AI) is applied in traffic management, demand prediction and resource optimization, allowing a smarter and more efficient administration of urban services. Finally, blockchain is used to secure transactions and data, improve transparency, and reduce fraud in urban services, providing an additional layer of security and trust in city systems.

A smart city integrates a variety of technologies and approaches to improve mobility and other urban services. The combination of digital infrastructure, IoT sensors, Big Data, citizen participation, 5G networks, open data platforms, artificial intelligence and blockchain makes it possible to create a more efficient, sustainable and livable urban environment. With the continuous advancement of these technologies and their implementation in cities, we can expect a significant improvement in the quality of life of citizens and in the efficiency of urban services.

Urban mobility and smart transport

Urban mobility in smart cities focuses on the integration of different modes of transport and the use of advanced technologies to improve efficiency and reduce environmental impact.

Smart public transport systems are central to this approach. The implementation of electric buses and trams helps reduce emissions and improve air quality. Not only do these electric vehicles decrease dependence on fossil fuels, but they also provide a quieter and cleaner means of transportation. In addition, real-time traffic management systems use sensors and artificial intelligence (AI) algorithms to optimize traffic flow and reduce congestion. These systems analyze data in real-time to adjust traffic lights and divert traffic as needed, improving the efficiency of urban transportation. On-demand public transport adjusts its routes and schedules based on demand in real time, improving efficiency and user service. This allows for greater flexibility and responsiveness to passenger needs.

Shared mobility is another key component of urban mobility in smart cities. Carsharing and ridesharing platforms allow users to share vehicles, reducing the need for car ownership and decreasing congestion. These platforms not only optimize vehicle usage, but also help reduce traffic and emissions. Bikesharing and scootersharing programs offer sustainable mobility options for short distances, promoting the use of greener and healthier means of transportation. In addition, fleets of shared autonomous vehicles can be used on demand, combining the convenience of private transport with the efficiency of public transport. These autonomous vehicles can operate continuously, optimizing routes and reducing the need for parking in congested areas.

The integration of transport modes is essential for efficient urban mobility. Mobility as a Service (MaaS) is an approach that integrates all modes of transportation into a single application, allowing users to plan, book, and pay for their trips efficiently. MaaS facilitates the transition between different types of transportation, from bike sharing to buses and trains, improving the user experience and promoting the use of multiple modes of transportation. Mobility hubs are hubs that combine different transport services, such as shared bikes, electric vehicle charging stations and bus stops, facilitating transitions between transport modes. These hubs provide an integrated infrastructure that supports more seamless and connected mobility.

Examples of smart cities that are implementing advanced urban mobility solutions include Singapore, Barcelona, and Helsinki. Singapore uses an AI-based traffic management system and sensors to optimize traffic flow and reduce congestion. In addition, it has implemented autonomous vehicles in specific areas and has an advanced integrated public transport system. Barcelona has implemented shared mobility solutions and created a network of "superislands" that prioritize pedestrians and cyclists, reducing car traffic in certain areas. These superislands improve the quality of life by reducing pollution and noise, and foster a friendlier environment for pedestrians and cyclists. Helsinki is pioneering MaaS with the Whim app, which allows users to access a variety of public and private transport services through a single platform. This app makes it easy to plan and pay for trips, integrating different modes of transportation and offering a complete and convenient mobility solution.

Urban mobility in smart cities is based on the integration of advanced technologies and the coordination of different modes of transport. Through smart public transport systems, shared mobility, integration of transport modes and examples of innovative cities, efficiency can be improved, environmental impact reduced and the quality of life of citizens increased. With the adoption of these practices and technologies, cities can move towards a more sustainable and efficient mobility future.

Traffic Management & Urban Planning

Efficient traffic management and smart urban planning are essential for sustainable mobility in smart cities.

Smart traffic management systems are critical to optimizing vehicle flow and reducing congestion. Traffic sensors and cameras collect real-time data on traffic flow, enabling traffic signal optimization and incident management. This data allows authorities to monitor and respond quickly to changes in traffic conditions, improving the efficiency of the transportation system. Adaptive Sign Control uses artificial intelligence (AI) algorithms to adjust traffic signals based on current conditions, reducing congestion and improving vehicle flow. This dynamic approach ensures that traffic flows more evenly, minimizing waiting times and reducing emissions. User information platforms provide real-time information on traffic, transport options and travel times, helping users to plan their routes more efficiently. These applications allow drivers and passengers to make informed decisions about their journeys, optimising their time and reducing the burden on transport infrastructures.

Sustainable urban planning is another key pillar for achieving efficient and green mobility in smart cities. Low emission zones are areas where access to high-emission vehicles is restricted, incentivizing the use of electric vehicles and sustainable modes of transport. These zones help improve air quality and encourage the use of cleaner transportation alternatives. The design of public and green spaces that prioritize pedestrian spaces and green areas improves the quality of life and promotes active modes of transportation such as walking and cycling. Not only do these spaces provide a more pleasant and healthy environment for residents, but they also reduce dependence on the car. Transit-oriented development (TOD) focuses on the development of housing and services around public transport nodes, reducing the need for car use and encouraging the use of public transport. This approach ensures that services and amenities are accessible to everyone, regardless of their reliance on private transportation.

Examples of smart urban planning can be found in cities such as Amsterdam, Stockholm and Copenhagen. Amsterdam has developed an extensive system of bike lanes and implemented policies that favor the use of bicycles and public transport. The city has also developed pedestrian areas and restricted vehicle access in certain areas, promoting a cleaner and safer urban environment for pedestrians and cyclists. Stockholm, known for its urban toll system, has significantly reduced congestion and emissions in the city centre. The city has also implemented low-emission zones and improved its public transport network, offering residents different cleaner and more efficient alternatives. Copenhagen has prioritised cycling infrastructure, with a network of cycle lanes covering the entire city. In addition, Copenhagen has developed pedestrian areas and implemented policies to reduce car use, promoting a more active and sustainable lifestyle.

Efficient traffic management and smart urban planning are crucial to creating sustainable and liveable cities. Through the implementation of advanced traffic management systems, the design of public and green spaces, and the development of transit-oriented transport infrastructure, cities can improve urban mobility, reduce environmental impact, and increase the quality of life of their citizens. With successful examples such as Amsterdam, Stockholm and Copenhagen, it is clear that these strategies can significantly transform the urban environment and make our cities more sustainable and efficient.

Sustainable mobility technologies in smart cities

Sustainable mobility technologies are a key component of smart cities, improving efficiency and reducing the environmental impact of transport.

Electric vehicles and charging stations are essential to this transformation. Deploying fast-charging networks in strategic locations facilitates the use of electric vehicles (EVs) by enabling fast and convenient charging. These stations, located in key locations such as highways and city centers, ensure that drivers can recharge their vehicles in minutes instead of hours, significantly improving the viability of EVs. Integrating these charging stations with renewables, equipping them with solar panels and energy storage systems, reduces dependence on the electricity grid and lowers carbon emissions. In addition, cargo management platforms optimize the use of stations, managing demand and reducing operating costs. These systems ensure that the charging infrastructure is used efficiently, avoiding overloads and maximizing availability for users.

Autonomous transport systems also play a crucial role in sustainable mobility. Fleets of autonomous buses operating on fixed or on-demand routes improve efficiency and reduce operating costs by eliminating the need for human drivers. These buses can operate continuously, optimizing routes and schedules according to actual demand. Autonomous taxi services offer a convenient and sustainable alternative to using the private car, providing on-demand transportation with lower operating costs and no direct emissions. Integration with smart infrastructure allows these autonomous transportation systems to communicate with road infrastructure to optimize traffic flow and improve safety. This communication helps to coordinate traffic, reduce congestion and improve road safety.

Open data platforms and advanced analytics are critical to the efficient management of smart cities. Using sensors and IoT devices to collect real-time data on traffic, air quality, and other urban parameters provides a rich and up-to-date database that can be used to make informed decisions. Open data platforms allow public access to this data, encouraging innovation and citizen participation. By providing accessible data, citizens and developers can create new applications and services that improve the quality of urban life. Advanced analytics uses big data and AI algorithms to analyze this data, helping authorities make informed decisions about traffic management and urban planning. This allows for a more efficient and proactive management of urban resources, improving the quality of life and reducing environmental impact.

Examples of sustainable mobility technologies include the Tesla Supercharger Network, Waymo, and Barcelona Smart City. The Tesla Supercharger Network is a network of fast-charging stations for Tesla vehicles, equipped with fast-charging technologies and, in some cases, solar panels, making it easy to recharge EVs quickly and sustainably. Waymo develops autonomous vehicles and offers autonomous taxi services in several cities in the United States, demonstrating the potential of autonomous mobility to reduce congestion and emissions. Barcelona Smart City uses sensors and open data platforms to manage traffic, optimise energy use and improve air quality, showing how the integration of advanced technologies can transform urban management.

Sustainable mobility technologies are essential for smart cities. The combination of electric vehicles, autonomous transport systems and advanced data platforms allows for more efficient management of urban transport, reducing environmental impact and improving the quality of life of citizens. With the implementation of these technologies, cities can move towards a more sustainable and efficient future.

Impact of mobility on smart cities

Sustainable mobility not only has environmental benefits, but also social and economic ones. Smart cities are taking advantage of these advantages to improve the quality of life of their inhabitants and foster economic development.

The social benefits of sustainable mobility are numerous. Improving air quality is one of the most important, as reducing vehicle emissions improves air quality, reducing respiratory and cardiovascular diseases. This not only improves the health of citizens, but also reduces the burden on public health systems. Accessibility and equity are other key benefits, as sustainable mobility improves access to economic opportunities and services for all citizens, including those in disadvantaged communities. This can reduce inequalities and promote greater social inclusion. In addition, promoting active modes of transportation, such as walking and cycling, improves the physical and mental health of citizens. These modes of transport are not only sustainable, but also encourage a more active and healthy lifestyle.

The economic benefits of sustainable mobility are also significant. Reducing operating costs is one of the main ones, as the improved efficiency and lower maintenance costs of electric vehicles (EVs) and intelligent transport systems reduce operating costs for businesses and public administrations. This can free up resources to be used in other important areas. Investment in sustainable mobility technologies and associated infrastructure creates jobs and fosters economic growth. The construction and maintenance of infrastructure, as well as the development of new technologies, generate employment opportunities and promote innovation. In addition, improving air quality and promoting active modes of transport reduce costs associated with pollution-related diseases and sedentary lifestyles, generating significant public health savings.

Examples of the social and economic impact of sustainable mobility can be found in several cities around the world. In Curitiba, Brazil, the implementation of an efficient and accessible public transport system has improved the quality of life and fostered economic development in the city. Curitiba's bus rapid transit (BRT) system is a model for other cities to follow, demonstrating how well-planned transport infrastructure can transform a city. In Zurich, Switzerland, the focus on public transport and cycling infrastructure has reduced congestion, improved air quality and increased citizen mobility, with significant public health benefits. Zurich has shown that a combination of efficient public transport and robust cycling infrastructure can improve urban mobility and quality of life. Seoul, South Korea, has implemented smart mobility solutions that have improved transportation efficiency, reduced emissions, and fostered economic growth. The integration of advanced technologies in traffic management and public transport has made Seoul a leader in sustainable mobility.

Sustainable mobility offers numerous social and economic benefits. By improving air quality, promoting equity and accessibility, and encouraging active modes of transportation, cities can improve the health and well-being of their citizens. At the same time, reduced operating costs, economic growth and public health savings demonstrate that sustainable mobility is a smart investment for the future. With successful examples such as Curitiba, Zurich, and Seoul, it is evident that smart cities can leverage sustainable mobility technologies to create healthier, more equitable, and more prosperous urban environments.

Challenges and barriers to implementation

Despite the numerous benefits, the implementation of sustainable mobility solutions in smart cities faces several challenges and barriers that must be overcome.

Startup cost and financing are big hurdles. The development of sustainable mobility infrastructure requires significant investments that can be a barrier for many cities. These investments include building charging stations, implementing intelligent transportation systems, and upgrading existing infrastructure. Finding sustainable financing models that involve both the public and private sectors is crucial for long-term success. Public-private partnerships (PPPs) can provide the resources and expertise needed to develop and maintain these infrastructures, sharing the risks and benefits among the parties involved.

Public acceptance and behavior change also pose significant challenges. Citizens may resist adopting new technologies and modes of transportation, especially if they are used to the use of private cars. This resistance may be due to a lack of information, the perception of inconvenience, or simply the inertia of habitual behavior. It is essential to educate and raise awareness among citizens about the benefits of sustainable mobility and to encourage changes in behaviour. Awareness campaigns can highlight the environmental, economic and health benefits associated with sustainable mobility, incentivising citizens to try and adopt new transport options.

The integration of technologies is another key challenge. Ensuring that different systems and technologies can work together efficiently is critical to the success of sustainable mobility. Interoperability between data platforms, transport systems, and IoT devices is essential to creating a cohesive and efficient network. In addition, protecting citizens' data and ensuring the security of smart systems is critical to gaining public trust. Smart mobility systems must be designed with robust security measures in place to protect against unauthorized access and misuse of personal data.

Regulations and standards must also be adapted to support sustainable mobility. It is necessary to develop regulatory frameworks that support the implementation of sustainable mobility technologies and protect the interests of citizens. These regulations should address issues such as security, privacy, interoperability, and fairness. In addition, regulations must be flexible and adapt to the rapid evolution of technologies. This involves creating policies that can evolve over time and respond to changes in the technology and market landscape.

Examples of overcoming challenges in sustainable mobility can be found in cities such as Amsterdam and Singapore. Amsterdam has overcome funding challenges through public-private partnerships and fostered public acceptance through awareness and education campaigns. The city has implemented programs highlighting the benefits of cycling and public transportation, and has worked with private companies to develop sustainable transportation infrastructure. Singapore has developed advanced regulatory frameworks and implemented security and privacy technologies to protect citizens' data and ensure the interoperability of its smart mobility systems. The city-state has invested in smart transportation solutions and created a regulatory environment that facilitates innovation and the adoption of new technologies.

Although there are significant challenges to the implementation of sustainable mobility solutions in smart cities, these can be overcome with appropriate strategies. Public-private sector collaboration, public education and awareness, efficient technological integration and the development of flexible regulatory frameworks are essential to create sustainable transport systems that improve quality of life and protect the environment. With successful examples such as Amsterdam and Singapore, it is clear that it is possible to overcome these challenges and move towards a more sustainable and efficient mobility future.

Future of mobility in smart cities

As technologies continue to evolve and smart cities adopt more integrated and sustainable approaches, the future prospects for urban mobility are promising.

Technological advances play a crucial role in improving urban mobility. The use of artificial intelligence (AI) and machine learning will improve traffic management, urban planning, and transportation efficiency. These technologies can analyze large volumes of data in real time, allowing for better decision-making and optimization of resources. The expansion of the Internet of Things (IoT) will enable greater connectivity and optimization of transportation systems, improving efficiency and convenience. IoT devices can collect data from sensors installed in vehicles and infrastructure, providing valuable insights to manage traffic and improve the user experience. Using blockchain to secure transactions and data can improve transparency and reduce fraud in mobility services. This technology can ensure data integrity and security, facilitating trust in transportation systems.

Innovative mobility models are transforming the way citizens plan and deliver their trips. The expansion of Mobility as a Service (MaaS) will transform the way citizens plan and deliver their journeys, integrating all modes of transport into a single platform. MaaS makes it easier to plan, book, and pay for trips, offering a more integrated and convenient user experience. Autonomous and shared vehicles will offer new mobility options, reducing the need for car ownership and improving the efficiency of urban transport. These vehicles can operate continuously, optimizing routes and reducing congestion.

Sustainable urban planning approaches are also essential for the future of urban mobility. Transit-oriented development (TOD) focuses on development around public transport nodes, encouraging the use of public transport and reducing dependence on the car. This approach creates more accessible and connected communities, making it easier to access services and jobs without the need for a private car. The expansion of low-emission zones and pedestrian areas will improve air quality and promote active modes of transportation. Not only do these areas reduce emissions, but they also create healthier and more attractive environments for residents.

Public policies and collaboration are key to supporting the transition to more sustainable urban mobility. The development of flexible and adaptive regulations will foster innovation and the adoption of sustainable mobility solutions. Regulations must be able to adapt quickly to technological advances and changing market needs. The international collaboration will allow the sharing of best practices and technologies, accelerating the transition to sustainable mobility at a global level. Cooperation between cities and countries can facilitate the exchange of knowledge and resources, supporting the development of more effective and efficient solutions.

The future of urban mobility is bright, thanks to technological advances, innovative mobility models, sustainable urban planning approaches and adaptive public policies. With the combination of AI, IoT, and blockchain, along with the expansion of MaaS and shared autonomous vehicles, cities can significantly improve the efficiency and sustainability of their transportation systems. At the same time, transit-oriented urban planning and the creation of low-emission zones will promote a healthier and more accessible environment. With international collaboration and flexible regulations, the path to sustainable urban mobility is well underway.

Conclusion

Smart cities are at the heart of the sustainable mobility revolution. By integrating advanced technologies, innovative urban planning approaches, and sustainable mobility models, these cities are improving the quality of life of their inhabitants and reducing their environmental impact. Despite the challenges, the opportunities to move towards a more efficient, cleaner and equitable transport system are immense.

The examples presented in this chapter demonstrate that, with the right approach, it is possible to transform urban mobility and achieve sustainability goals. As technologies continue to evolve and cities adopt more integrated approaches, the future of sustainable mobility in smart cities is promising. With the commitment and collaboration of all sectors of society, we can build a transportation system that not only meets our mobility needs, but also protects our planet for future generations.

Chapter 7: Citizenship and Education in Sustainable Mobility

The transition to sustainable mobility depends not only on technological advances and effective public policies, but also on the participation of citizens and appropriate education that fosters a culture of sustainability. This chapter explores the importance of citizen participation and education in promoting sustainable mobility, highlighting strategies, initiatives and case studies that have proven successful in various parts of the world.

Awareness campaigns and public education

Public awareness and education campaigns are critical to inform citizens about the benefits of sustainable mobility and motivate changes in behavior.

Effective strategies for awareness campaigns include the use of mass media, community events, and collaboration with schools and universities. Campaigns on television, radio and social networks can reach a wide and diverse audience. Clear and persuasive messages about the benefits of sustainable mobility, such as reducing emissions and improving public health, can raise awareness and motivate changes in behaviour. Mass media makes it possible to disseminate information quickly and reach different segments of the population, which is essential to create a broad and meaningful movement towards more sustainable mobility practices.

Hosting community events such as car-free days, sustainable mobility fairs, and educational workshops can provide hands-on, hands-on experiences. These events allow citizens to try new forms of transportation, learn about their benefits, and discuss concerns and questions with experts. Car-free days, for example, temporarily transform streets into car-free spaces, offering residents the opportunity to experience the city in a different way and see the immediate benefits of reduced motorized traffic. Educational fairs and workshops can provide detailed information and enable direct interaction with sustainable mobility technologies and practices, fostering deeper understanding and long-term engagement.

Engaging educational institutions in awareness campaigns can have a lasting impact. Educational programs that integrate sustainable mobility into the school curriculum can educate young people about the importance of these practices from an early age. Collaborating with schools and universities not only helps to form sustainable habits in students, but can also influence their families and communities, expanding the reach of the campaign. School activities can include class projects on sustainable transport, excursions that use non-motorised modes of transport and competitions that promote the use of bicycles or walks.

Examples of successful campaigns demonstrate how these strategies can have a significant impact. "Bike to Work Day" in the United States is an annual event that encourages citizens to use bicycles to commute to work. The campaign includes guided tours, refreshment stations and prizes for participants, and has been successful in increasing cycling and raising awareness of the benefits of urban cycling. This type of event not only promotes the use of the bicycle, but also creates a community of cyclists who can share experiences and support each other.

"European Mobility Week" is an initiative of the European Commission that promotes sustainable urban mobility through events and activities in cities across Europe. The campaign includes mobility challenges, workshops and conferences, and has managed to increase awareness and citizen participation in sustainable mobility. This week of activities provides a platform for cities to share their best practices and motivate residents to try new forms of sustainable transport.

"Recreational Bike Paths" in Latin America are another example of success in sustainable mobility campaigns. Cities such as Bogota and Mexico City close certain streets to car traffic during weekends, allowing citizens to walk, bike, and enjoy outdoor activities. These initiatives encourage the use of bicycles and raise awareness about the benefits of car-free urban spaces. Recreational bike lanes not only promote exercise and active transportation, but also temporarily transform urban space, demonstrating the potential for a less car-dependent environment.

Public awareness and education campaigns are essential to promote sustainable mobility. Using mass media, organizing community events, and engaging educational institutions, these campaigns can inform and motivate citizens, encouraging a change in behavior towards more sustainable transportation practices. Examples such as "Bike to Work Day", "European Mobility Week" and

"Recreational Cycle Paths" show how these strategies can have a positive and lasting impact on society.

Community Participation in Transportation Planning

Citizen participation in transport planning is crucial to ensure that sustainable mobility solutions respond to the needs and preferences of the community.

Community engagement mechanisms are diverse and effective in engaging citizens in the transportation planning process. Organising public consultations and hearings allows citizens to express their opinions and concerns about transport projects. These interactions can provide valuable information to planners and ensure that decisions are inclusive and representative. Public consultations are a forum where citizens can discuss directly with decision-makers, ensuring that community voices are heard and considered in the design of policies and projects.

Conducting surveys and opinion studies can help to better understand the needs and preferences of citizens. The results of these surveys can guide the design and implementation of transport policies and projects, ensuring that they respond to the real demands of the community. Surveys can cover a wide range of topics, from transportation preferences to environmental concerns, providing a solid foundation for informed decision-making.

Forming committees and working groups with community representatives can facilitate an ongoing and constructive dialogue between citizens and planners. These groups can be involved in all stages of the planning process, from conception to implementation and evaluation. By including representatives from diverse parts of the community, it ensures that mobility solutions are equitable and reflect the needs of all citizens. Committees can provide constant feedback and act as liaisons between planners and the community at large.

Examples of community participation in transport planning can be found in various parts of the world. In Spain, many cities have implemented the Sustainable Urban Mobility Plan (SUMP), which includes citizen participation processes to develop and evaluate sustainable mobility strategies. Citizens participate in workshops, surveys and meetings, ensuring that mobility solutions respond to their needs. This participatory approach has led to the development of strategies that are not only effective, but also have the support of the community.

In Porto Alegre, Brazil, Participatory Budgeting allows citizens to decide how a portion of the municipal budget is spent. Sustainable mobility is one of the priority areas, and citizens can propose and vote for projects that improve public transport, cycling infrastructure and other mobility solutions. Not only does this process empower citizens, but it also ensures that funds are used on projects that have significant community support.

In the United Kingdom, the "Living Streets" initiative temporarily transforms residential streets into spaces without car traffic, allowing residents to experience and discuss the potential for permanent changes. Citizen feedback is used to inform urban redesign projects that promote sustainable mobility. This initiative allows residents to visualize the benefits of reduced traffic and actively participate in creating more livable and sustainable urban environments.

Citizen participation is essential for the success of sustainable mobility solutions in smart cities. Using mechanisms such as public consultations, surveys, and community committees, planners can ensure that decisions reflect the needs and preferences of the community. Successful examples of community engagement, such as the SUMP in Spain, the Participatory Budget in Porto Alegre and "Living Streets" in the United Kingdom, demonstrate how citizen inclusion can lead to more effective and supported mobility solutions. With robust citizen participation, cities can move towards a more sustainable and equitable mobility future.

Educational programs and training

Education and training are essential to foster a culture of sustainable mobility and empower citizens and professionals to implement and maintain sustainable transport solutions.

Initiatives in schools and communities are critical to instilling sustainable values and practices from an early age and among citizens. Integrating sustainable mobility into the school curriculum can educate students about the importance of sustainable practices and prepare them to be advocates and users of these solutions. Activities such as science projects, field trips, and "safe pathways to school" programs can have a lasting impact, teaching young people how to value and use sustainable modes of transportation in their daily lives. Community workshops and courses can provide citizens with practical knowledge and skills. These programs can address topics such as efficient driving, bike maintenance, using public transportation, and planning sustainable trips. By equipping citizens with these skills, they are empowered to make more informed and sustainable transportation choices.

Training and supporting community ambassadors who promote sustainable mobility in their neighborhoods can increase awareness and participation. These ambassadors can organize events, provide information, and serve as liaisons between citizens and planners. By having community members act as local advocates, greater acceptance and adoption of sustainable practices can be fostered, as citizens can see tangible examples of how these practices benefit their community.

Training for transport professionals is equally important to ensure that sustainable mobility solutions are effectively implemented and maintained. Offering training and certification courses for transport professionals can ensure that they have the skills and knowledge needed to implement and maintain sustainable mobility solutions. These programmes may cover topics such as cycling infrastructure design, public transport management and sustainable urban planning. Proper training ensures that professionals can apply the latest best practices and technologies in their projects.

Facilitating exchange and learning programmes between cities and regions can allow transport professionals to share experiences, learn from best practices and develop new ideas and solutions. Conferences, workshops and study visits are examples of these initiatives. By sharing knowledge and experiences, professionals can adapt and improve their sustainable mobility strategies, benefiting from the successes and failures of others.

Examples of educational programs and training demonstrate how these initiatives can have a significant impact. "Safe Routes to School" in the United States is a national program that promotes the use of active and sustainable modes of transportation to get to school. It offers educational resources, training for local coordinators, and funding for safe and accessible infrastructure. This program has been successful in increasing safety and participation in sustainable modes of transportation among students.

"Copenhagenize Design Company" in Denmark offers training courses and workshops on urban design and cycling mobility. It aims to educate transport professionals and urban planners on best practices in promoting urban cycling. By providing specialized training, this company helps develop more cyclist-friendly and sustainable cities.

"Mobility Academy" in Switzerland offers training courses and certification programs for transport professionals. Topics range from public transport planning to the implementation of shared and sustainable mobility solutions. This academy provides professionals with the tools and knowledge necessary to lead the transition to more sustainable transport systems.

Education and training are crucial to foster a culture of sustainable mobility. Through educational programs in schools and communities, and specialized training for transportation professionals, the skills and knowledge needed to implement and maintain sustainable mobility solutions can be developed. Examples such as "Safe Routes to School", "Copenhagenize Design Company" and "Mobility Academy" show how education and training can have a positive and lasting impact on urban mobility. With proper education and training, cities can move towards a more sustainable and equitable future.

Impact of citizen participation and education

Citizen participation and education can have a significant impact on promoting sustainable mobility, improving the adoption of sustainable transport solutions and fostering a culture of sustainability.

The benefits of citizen participation are numerous and essential for the success of sustainable mobility policies. Citizen participation in transport planning can increase the legitimacy and acceptance of policies and projects, ensuring that they respond to the needs and preferences of the community. When citizens feel heard and involved in the decision-making process, they are more likely to support and adopt the proposed initiatives. In addition, citizen feedback can provide valuable information that improves the quality and relevance of sustainable mobility solutions, ensuring that they are effective and viable. This information can include specific details about local needs and barriers residents face in their daily lives.

Citizen participation can also foster a culture of sustainability, motivating citizens to adopt sustainable practices and be active advocates for sustainable mobility. By involving citizens in the planning and execution of transportation projects, a sense of ownership and shared responsibility is created. This commitment can extend beyond transport, influencing other aspects of urban life and promoting a more sustainable attitude in general.

The benefits of education and training are equally crucial. Education and training can provide citizens and transport professionals with the skills and knowledge needed to implement and maintain sustainable mobility solutions. Educational programs can teach students and the community about the importance of sustainable mobility, as well as practical techniques for adopting these practices in their daily lives. Education can also increase awareness of and commitment to sustainable mobility, motivating citizens to adopt sustainable practices and actively participate in mobility initiatives. This increase in awareness can lead to increased demand for sustainable mobility solutions and increased pressure on decision-makers to implement changes.

Education and training can also foster innovation and creativity, enabling citizens and transport professionals to develop new ideas and solutions for sustainable mobility. By providing an environment where new ideas are valued and experimentation is supported, cities can benefit from innovative solutions that improve the efficiency and sustainability of urban transport.

Examples of the positive impact of citizen participation and education on sustainable mobility can be seen in several cities around the world. In Portland, Oregon, United States, the city has implemented an extensive citizen engagement and education program to promote sustainable mobility. Citizens have been actively involved in transportation planning, and the city has seen a significant increase in the use of public transportation, cycling, and active modes of transportation. This inclusive approach has resulted in solutions that are more suited to local needs and that have broad community support.

Vancouver, Canada, has developed educational and training programs for citizens and transportation professionals. These programs have increased awareness of and commitment to sustainable mobility and encouraged the adoption of sustainable practices throughout the city. Training professionals ensures that the solutions implemented are of high quality and based on best practices, while community education promotes wider adoption of these practices.

Freiburg, Germany, is known for its focus on sustainability and has actively involved citizens in transport planning. The city has implemented educational programs in schools and the community, achieving a high adoption of sustainable mobility, with significant use of public transportation and cycling. Freiburg demonstrates how the combination of citizen participation and education can create a culture of sustainability that benefits the entire community.

Citizen participation and education are essential for the promotion of sustainable mobility. Through the active inclusion of citizens in planning and the provision of appropriate education and training, cities can develop transport solutions that are effective, accepted and sustainable. Examples such as Portland, Vancouver and Freiburg show how these approaches can have a positive and lasting impact on urban mobility.

Challenges and barriers to citizen participation and education

Despite the benefits, the promotion of citizen participation and education in sustainable mobility faces several challenges and barriers that need to be addressed.

Challenges to citizen participation include lack of resources and funding, inequalities in participation, and resistance to change. Implementing citizen participation processes requires resources and funding, which can be limited in many cities and communities. Without adequate funding, it is difficult to organize events, conduct surveys, and ensure that all voices are heard. Social, economic, and geographic inequalities can affect citizen participation, leaving out certain groups and communities. This means that some people's views and needs are not adequately considered, which can lead to mobility solutions that are not inclusive or equitable. Resistance to change and lack of interest can be barriers to citizen participation in transport planning. Many people may be reluctant to change their transportation habits or participate in planning processes that they perceive as complicated or irrelevant to their daily lives.

The challenges of education and training are also significant. Access to educational and training programs may be limited for certain groups and communities, especially in rural and disadvantaged areas. Without adequate access, many people cannot benefit from sustainable mobility education or develop the skills needed to adopt sustainable practices. The relevance and quality of educational and training programs can vary, affecting their effectiveness in promoting sustainable mobility. Programs that are poorly designed or do not address the specific needs of the community may fail to generate the desired impact. Educational and training programs need to be tailored to the needs and preferences of different audiences, which can be challenging in practice. Creating content that is accessible and relevant to diverse populations requires time, resources, and a deep understanding of different communities.

Examples of overcoming challenges in promoting citizen participation and education in sustainable mobility can be found in several cities. Bogotá, Colombia, has implemented citizen engagement programs that include disadvantaged communities and has provided funding to ensure that all groups can actively participate. The city has also developed educational programs in collaboration with schools and community organizations. These efforts have ensured that a wide range of voices are heard and that citizens are informed and engaged with sustainable mobility initiatives.

Seoul, South Korea, has developed educational and training programs accessible to all citizens, including online courses and free educational resources. The city has also worked to ensure that programs are relevant and of high quality, using continuous feedback to improve. By offering flexible and accessible educational options, Seoul has been able to engage a wide audience and encourage greater adoption of sustainable practices.

Melbourne, Australia, has implemented citizen engagement programs that foster interest and motivation through incentives and rewards. The city has also developed educational programs that cater to different audiences, ensuring they are inclusive and accessible. These programs not only educate citizens, but also motivate them to actively participate in the planning and adoption of sustainable mobility solutions.

Although there are significant challenges for the promotion of citizen participation and education in sustainable mobility, these can be overcome with appropriate strategies. Including disadvantaged communities, using adequate funding, creating accessible and relevant educational programs, and adapting content for diverse audiences are crucial steps. Examples such as Bogota, Seoul and Melbourne show how these cities have addressed and overcome these challenges, making a positive impact on sustainable urban mobility.

Future of participation and education in sustainable mobility

As cities and communities continue to move towards sustainable mobility, citizen engagement and education will continue to be crucial components for success.

Technological and digital advances are facilitating citizen participation and access to education in sustainable mobility. The use of digital participation platforms and mobile applications can facilitate citizen participation, allowing citizens to provide feedback and participate in transport planning in a more convenient and accessible way. These digital tools can be designed to collect opinions, suggestions and concerns from citizens, ensuring that their voices are heard and considered in the decision-making process. The technology also allows for more efficient and direct communication between planners and the community, streamlining the flow of information and increasing transparency.

The expansion of online educational resources and online courses can increase access to education and training in sustainable mobility, especially for rural and disadvantaged communities. Online courses offer flexibility in terms of time and place, which is essential for people with busy schedules or who live in areas where face-to-face educational programs are not available. These resources can include instructional videos, interactive tutorials, and discussion forums that allow participants to learn at their own pace and connect with others interested in sustainable mobility.

Inclusive and equitable approaches are essential to ensure that all citizens have the opportunity to participate in transport planning and access education on sustainable mobility. Developing participatory approaches that include all groups and communities ensures that the voices of all citizens are heard and represented in transportation planning. This involves implementing specific strategies to engage marginalized or traditionally underrepresented groups, such as low-income people, rural communities, and ethnic minorities. The active inclusion of these voices is crucial to developing mobility solutions that are truly equitable and accessible to all.

Ensuring that educational and training programmes are accessible and relevant to all citizens, regardless of their socio-economic, geographical or demographic situation, is another fundamental aspect. Programs should be designed taking into account the diverse needs and contexts of the participants. This may include adapting educational materials for different levels of knowledge, providing resources in multiple languages, and creating specific programs for different age groups and abilities.

Collaboration and cooperation are vital for the success of citizen participation and education in sustainable mobility. Fostering public-private partnerships can be an effective way to finance and support citizen engagement and education programs, ensuring they have the necessary resources and sustainability. These partnerships can combine the resources and expertise of the public and private sectors to develop more robust and effective programs. International cooperation and the sharing of best practices and resources between cities and countries can also be very beneficial. Facilitating international cooperation allows communities to learn from each other and improve their approaches to participation and education. Sharing experiences and knowledge across global networks can accelerate the adoption of innovative and effective solutions in sustainable mobility.

As cities and communities move towards sustainable mobility, citizen engagement and education will continue to be crucial components. Technological and digital advances, inclusive and equitable approaches, and collaboration and cooperation are essential to ensure that these initiatives are effective and sustainable. By leveraging digital technologies, developing inclusive approaches, and fostering collaboration, cities can significantly improve citizen engagement and education in sustainable mobility, benefiting the entire community.

Conclusion

Citizen participation and education are essential components for the promotion of sustainable mobility. Through awareness campaigns, community engagement, educational programs, and training, we can foster a culture of sustainability and empower citizens and transportation professionals to implement and maintain sustainable mobility solutions.

Despite the challenges, the opportunities to advance citizen participation and education are immense. With the right approach, we can increase awareness, commitment and adoption of sustainable practices, ensuring a cleaner, more efficient and equitable mobility future for all.

This chapter has explored the importance of citizen participation and education in sustainable mobility, providing examples of success and effective strategies. In the following chapters, we will continue to explore the technologies and policies that are shaping the future of mobility, with the aim of providing a comprehensive and compelling understanding of how we can build a more sustainable and equitable future for all.

Chapter 8: Economic Benefits of Sustainable Mobility

Sustainable mobility not only has environmental and social benefits, but also significant economic impacts. From reducing operating costs to creating jobs and boosting economic development, sustainable mobility solutions offer numerous advantages that can contribute to long-term economic growth. In this chapter, we will explore in depth the economic benefits of sustainable mobility and how it can generate savings, create employment opportunities and foster sustainable economic growth.

Reduced operating costs

One of the most immediate benefits of sustainable mobility is the reduction of operating costs for both companies and individuals. This savings manifests itself in various ways, positively impacting the personal and business economy.

First, the fuel savings are significant. Electric vehicles (EVs) and sustainable modes of transportation, such as cycling and public transportation, consume less energy than internal combustion vehicles. EVs, in particular, are much more energy efficient and have significantly lower fuel costs. By relying on electricity, which is often cheaper than petrol or diesel, EV users can see a considerable reduction in their daily transport costs.

In addition to fuel savings, reduced maintenance is another important factor. EVs have fewer moving components and do not require oil changes, air filters, spark plugs, and other routine maintenance associated with internal combustion engines. This translates into lower maintenance and repair costs. By having simpler and more durable systems, EVs not only reduce the need for frequent workshop visits, but also extend the life of the vehicle, generating long-term savings.

Operational efficiency also improves with the adoption of sustainable mobility. Telematics and fleet management systems allow the optimization of routes and times, reducing fuel consumption and improving operational efficiency. These systems can analyze data in real-time to suggest the fastest and most efficient routes, minimizing the time vehicles spend on the road and reducing wear and tear and fuel use.

Another significant advantage is the reduction of traffic congestion. Implementing sustainable mobility solutions can reduce traffic congestion, improving efficiency and reducing travel times. This not only saves individuals time and money, but also improves the productivity of businesses. Less time in traffic means more time to perform productive tasks, thus increasing the overall efficiency of business operations.

A prominent example of operating cost reduction is found in the delivery company DPDgroup. This company has deployed a fleet of EVs in several European cities, significantly reducing their fuel and maintenance costs. In addition, route optimization has improved operational efficiency, allowing for faster and cheaper deliveries. The adoption of EVs has enabled DPDgroup to not only reduce its operating costs, but also improve its environmental impact, aligning with its customers' sustainability expectations.

Another example is the city of Los Angeles, which has converted part of its bus fleet to electric. This change has reduced operating costs in terms of fuel and maintenance. The savings obtained have been reinvested in the expansion and improvement of the public transport system, offering better services to citizens and promoting the use of sustainable transport. This initiative has not only improved the efficiency of the city's transport system, but has also contributed to the reduction of polluting emissions, improving urban air quality.

Sustainable mobility offers numerous economic benefits through reduced operating costs. Fuel and maintenance savings, along with optimised operational efficiency and reduced congestion, enable businesses and individuals to enjoy more economical and efficient transport. Examples such as DPDgroup and the City of Los Angeles demonstrate how the adoption of sustainable mobility solutions can transform both business operations and urban transportation systems, generating significant savings and contributing to a more sustainable future.

Economic growth and job creation

The transition to sustainable mobility can be a driver of economic growth and job creation, generating new opportunities in various sectors. This change not only benefits the environment, but also boosts the economy and improves the quality of life for many people.

The electric vehicle (EV) industry is one of the sectors that has benefited the most from this transition. The growing demand for EVs has driven job creation in vehicle and battery manufacturing. Companies like Tesla, Rivian, and Lucid Motors are expanding their production plants and hiring thousands of workers. EV manufacturing not only includes the production of automobiles, but also the creation of advanced batteries, which requires a specialized workforce in engineering and manufacturing. This industrial expansion is generating a wave of employment in regions where new plants and production lines are being established.

EV charging infrastructure development is also creating jobs in the installation, maintenance, and operation of charging stations. Companies like ChargePoint and EVBox are leading this growth, establishing networks of charging stations around the world. The construction and maintenance of these stations require skilled workers in various areas, from electrical engineering to customer service. This infrastructure is crucial to support the increase in EV adoption and ensure that drivers have convenient access to charging.

Public transport and shared mobility are also seeing a boom in job creation. Investment in sustainable public transportation, such as electric buses and trains, creates jobs in the manufacturing, operation, and maintenance of these systems. In addition, the construction of new lines and stations creates opportunities in the construction sector. Cities that invest in public transportation not only improve their infrastructure, but also create stable, long-term jobs that benefit the community.

Shared mobility services, such as carsharing, bikesharing, and scootersharing, are creating jobs in fleet management, customer service, and vehicle maintenance. Startups like Lime, Bird, and Zipcar are rapidly expanding their operations and hiring staff. These companies are developing innovative business models that require efficient logistics and constant attention to the maintenance and operability of their fleets, resulting in a variety of job opportunities.

Innovation and technological development are other important drivers of economic growth in the field of sustainable mobility. Investment in research and development (R+D) is generating jobs in engineering, technology and environmental sciences. Universities and research centers are playing a key role in the development of new technologies, such as more efficient batteries and energy management systems. These R+D efforts not only advance scientific knowledge, but also create highly skilled and well-paid jobs.

The transition to sustainable mobility has spawned an ecosystem of startups and startups that are innovating in areas such as electrification, artificial intelligence, and connectivity. These companies not only create jobs, but also attract investment and foster economic growth. The emergence of new technologies and business models is attracting investors and fostering competitiveness, which in turn generates more opportunities for employment and economic development.

Concrete examples of economic growth and job creation through sustainable mobility can be seen in different parts of the world. The construction and operation of Tesla's Gigafactory in Nevada, United States, has created thousands of jobs in the manufacture of batteries and electric vehicles. This plant has not only generated direct employment, but has also attracted suppliers and boosted economic development in the region. The presence of a facility of this magnitude has transformed the local economy, attracting professionals and encouraging the development of complementary infrastructure.

Another example is the public transportation system in Medellín, Colombia. The expansion of Medellín's public transportation system, which includes the metro, tram, and electric buses, has created numerous jobs and boosted local economic growth. Improved mobility has also increased accessibility to jobs and services for residents. This expansion has significantly improved the quality of life in the city, reducing congestion and improving transportation efficiency.

The transition to sustainable mobility is generating a wealth of economic opportunities. Job creation in the EV industry, the development of charging infrastructure, the expansion of public transportation, and technological innovation are driving economic growth and improving quality of life. Examples such as Tesla's Gigafactory and Medellín's public transport system demonstrate how sustainable mobility can be a driver of economic and social development.

Public Health Savings

Sustainable mobility can have a significant impact on public health, generating savings in medical costs and improving the quality of life of citizens. This approach not only contributes to the preservation of the environment, but also has direct benefits for people's health and well-being.

Reducing air pollution is one of the most immediate and palpable effects of sustainable mobility. The transition to electric vehicles (EVs) and non-motorized modes of transportation reduces greenhouse gas emissions, which contributes to mitigating climate change and improving air quality. Lower emissions of carbon dioxide and other harmful gases help curb global warming and its devastating effects on the planet. In addition, EVs and electric public transport emit fewer local pollutants, such as particulate matter and nitrogen oxides, which are harmful to respiratory and cardiovascular health. Decreasing these pollutants can reduce the incidence of diseases such as asthma, chronic bronchitis, and other respiratory conditions, as well as heart disease and stroke.

Promoting active modes of transportation, such as walking and cycling, has numerous public health benefits. Encouraging the use of bicycles and walking as active modes of transport can improve the physical and mental health of citizens. Not only are these modes of transportation a form of exercise, but they also help prevent chronic diseases such as obesity, diabetes, and heart disease. Regular physical activity is linked to better mental health, reducing the risk of depression and anxiety. In addition, developing safe and accessible infrastructure for pedestrians and cyclists can encourage greater physical activity and reduce the risk of road accidents. Well-designed bike lanes, safe crosswalks, and pedestrian zones can make these modes of transportation more attractive and safer for all citizens.

Concrete examples of how sustainable mobility can generate savings in public health are numerous and encouraging. The Vélib' bike-sharing program in Paris, France, has significantly increased bicycle use in the city, improving the health of residents and reducing air pollution. Studies have shown that the program has contributed to a decrease in rates of respiratory and cardiovascular diseases. By providing an accessible and sustainable transport alternative, Vélib' has helped transform urban mobility and generated substantial public health benefits.

In New York, United States, pedestrianization projects and the creation of bike lanes have improved air quality and encouraged active modes of transportation. The transformation of areas like Times Square into pedestrian zones has reduced local pollutant emissions and made streets more pedestrian- and cyclist-friendly. Not only have these projects improved the aesthetics and functionality of the city, but they have also led to significant savings in public health costs. By reducing exposure to pollutants and increasing physical activity, these initiatives have contributed to improving the health of New Yorkers.

Sustainable mobility offers multiple benefits for public health. Reducing air pollution and promoting active modes of transportation can prevent disease, improve mental and physical health, and lead to savings in medical costs. Examples such as the Vélib' program in Paris and pedestrianization projects in New York demonstrate how the implementation of sustainable mobility solutions can transform cities and improve the quality of life of their inhabitants. By continuing to promote and develop these initiatives, we can create healthier and more sustainable urban environments for all.

Economic incentives and return on investment

Economic incentives and return on investment are key aspects to encourage the adoption of sustainable mobility solutions and ensure their long-term viability. These incentives can motivate both individuals and companies to invest in technologies and practices that benefit the environment and the economy.

One of the most effective approaches to incentivize the adoption of sustainable mobility solutions is subsidies and tax exemptions. Governments can offer subsidies and tax breaks for the purchase of electric vehicles (EVs), electric bicycles, and other modes of sustainable transportation. These incentives can reduce the upfront cost, which is often a significant barrier, making these options more attractive to consumers. For example, subsidies can cover a portion of the cost of an EV, while tax breaks can reduce the taxes associated with owning and operating these vehicles, thereby increasing their affordability.

In addition to incentives for vehicle purchases, it is also crucial to encourage investment in infrastructure. Incentives for the installation of charging stations and cycling infrastructure can stimulate the necessary investments and facilitate the transition to sustainable mobility. Providing financial support for the creation of charging station networks can ensure that EV owners have convenient access to charging, removing one of the main barriers to mass adoption of these vehicles. Similarly, investing in bicycle infrastructure, such as bike lanes and safe parking lots, can incentivize more people to use bicycles as a means of daily transportation.

Financing and loan programs also play a vital role in promoting sustainable mobility. Low-interest loan programs for EV purchases and infrastructure installations can reduce financial barriers and encourage adoption of these technologies. By offering favourable terms, these loans can make sustainable mobility investments more accessible to individuals and businesses. In addition, the use of green bonds and other sustainable financing mechanisms can provide the necessary resources for sustainable mobility projects and ensure a long-term return on investment. Green bonds are debt instruments that finance projects with environmental benefits, such as EV charging infrastructure or electric public transport systems, attracting investors interested in supporting sustainability.

The return on investment for governments and companies investing in sustainable mobility solutions is significant. Savings in fuel and maintenance costs, as well as improved operational efficiency, can generate a considerable return on investment. EVs, for example, have lower operating costs due to their energy efficiency and lower maintenance needs compared to internal combustion vehicles. In addition, route optimization and congestion reduction can improve productivity and reduce operating expenses. In the long run, the economic benefits also manifest themselves in the form of higher tax revenues and lower public health costs. The improvement in air quality and public health resulting from emission reductions can reduce the burden on health systems, generating significant savings.

Concrete examples of economic incentives and return on investment can be observed in different parts of the world. Norway, for example, offers generous incentives for EV purchases, including tax breaks and subsidies. These incentives have significantly boosted EV adoption in the country and have proven to be a cost-effective investment, with benefits in terms of emissions reductions and improved air quality. Norway's strategy has made the country a global leader in EV adoption, demonstrating how well-designed incentives can transform the mobility market.

In Europe, green finance projects have played a crucial role in the development of charging infrastructure and electric public transport systems. The issuance of green bonds to finance sustainable mobility projects has provided the necessary resources for these initiatives, generating a significant return on investment in terms of savings in operating costs and long-term economic benefits. These projects have not only improved sustainable mobility infrastructure in several European cities, but have also encouraged investment in clean technologies and created jobs in sustainability-related sectors.

Economic incentives and return on investment are key to fostering the adoption of sustainable mobility solutions. Through subsidies, tax breaks, financing programs, and loans, governments can reduce financial barriers and make sustainable technologies more accessible. Economic benefits, such as savings in operating costs and long-term benefits for public health and the environment, underscore the importance of investing in sustainable mobility. Examples such as incentive schemes in Norway and green finance projects in Europe demonstrate how these strategies can transform urban mobility and generate significant benefits for society.

Examples of positive economic impact on sustainable mobility

To illustrate how sustainable mobility can generate significant economic benefits, let's present some outstanding examples from around the world. These cases demonstrate how investments in infrastructure and sustainable transport policies not only improve the quality of life of citizens, but also boost economic growth.

In Zurich, Switzerland, the city has developed one of the most efficient and sustainable public transport systems in the world, meeting the challenge of traffic congestion and the need to reduce emissions. The solution implemented has been a significant investment in a comprehensive network of electric trams, buses and trains, complemented by cycling and pedestrian infrastructure. In addition, Zurich has implemented traffic management systems and digital platforms to optimize urban mobility. The results have been impressive: the efficient public transport system has reduced traffic congestion, improved air quality and increased accessibility to employment and services. Investment in sustainable mobility has generated a significant economic return, with savings in operating costs and long-term benefits, showing that the right planning and investment can transform urban mobility and generate substantial economic advantages.

Amsterdam, in the Netherlands, is known for its cycling culture and has faced the challenge of promoting sustainable modes of transport in a densely populated city. The implemented solution includes the development of an extensive network of protected bike lanes, bikesharing programs and various incentives for the use of bicycles. In addition, the city has invested in safe and accessible cycling infrastructure, creating an environment that favors cycling as a primary mode of transportation. The results have been remarkably positive: the promotion of cycling has reduced dependence on cars, decreased emissions and improved public health. This cycling culture has also generated significant economic benefits, including the growth of the bicycle industry and cycling tourism, which has attracted numerous visitors and boosted the local economy.

Portland, Oregon, in the United States, has been a pioneer in the implementation of sustainable mobility solutions, facing the challenge of traffic congestion and the need to improve air quality. The city has invested in a comprehensive public transport network, cycling and pedestrian infrastructure, as well as carsharing and ridesharing programmes. In addition, Portland has implemented transit-oriented development (TOD) policies and encouraged citizen participation to ensure that mobility solutions respond to community needs. The results of these investments have been significant: reducing traffic congestion, improving air quality, and increasing accessibility to employment and services have transformed the city. Portland has seen remarkable economic growth, with job creation and the attraction of investments in sustainable technologies, demonstrating that cities can benefit economically by adopting sustainable mobility practices.

These examples demonstrate that sustainable mobility is not only good for the environment, but also offers tangible economic benefits. Investments in sustainable transport infrastructure, promoting active modes of transport and implementing favourable policies can lead to operational cost savings, improve public health and foster economic growth. Cities that adopt these practices can expect to see improvements in the quality of life for their residents and an increase in their long-term economic competitiveness. The experience of Zurich, Amsterdam and Portland shows that, with proper planning and a commitment to sustainability, it is possible to create efficient and economical transport systems that benefit everyone.

Economic challenges of sustainable mobility

Despite the economic benefits, the promotion of sustainable mobility faces several challenges and barriers that need to be addressed. These obstacles can make it difficult to implement sustainable solutions, but with proper planning and cooperation between different sectors, it is possible to overcome them.

One of the main challenges is the initial cost and financing. The development of sustainable mobility infrastructure requires significant investments that can be a barrier for many cities and communities. Building networks of electric vehicle (EV) charging stations, creating protected bike lanes, and expanding electric public transport are expensive projects that require a substantial financial commitment. Finding sustainable financing models that involve both the public and private sectors is crucial for long-term success. Collaboration between governments, private companies and investors can provide the necessary resources for these investments, ensuring that projects are viable and sustainable over time.

Another major challenge is public acceptance and behavior change. Citizens may resist adopting new technologies and modes of transportation, especially if they are used to the use of private cars. Changing deeply ingrained habits requires considerable effort in terms of education and awareness. It is essential to educate and raise awareness among citizens about the benefits of sustainable mobility and to encourage changes in behaviour. Information campaigns, educational programs in schools and communities, and the promotion of long-term benefits can help overcome this resistance and motivate people to adopt more sustainable transportation practices.

The integration of technologies is another key challenge. Ensuring that different systems and technologies can work together efficiently is critical to the success of sustainable mobility. Interoperability between transport systems, digital platforms and traffic management technologies is crucial to create a cohesive and efficient mobility ecosystem. In addition, data security and privacy are major concerns. Protecting citizens' data and ensuring the security of smart systems is critical to gaining public trust. Implementing robust cybersecurity measures and developing clear policies on data use and protection can help address these concerns.

Regulations and standards also play a crucial role in promoting sustainable mobility. It is necessary to develop regulatory frameworks that support the implementation of sustainable mobility technologies and protect the interests of citizens. Regulations must be flexible and adapt to the rapid evolution of technologies. This requires a proactive and collaborative approach between governments and industry to ensure that policies are effective and that technological advances can be adopted without unnecessary obstacles.

Concrete examples of overcoming challenges in sustainable mobility can be seen in various cities around the world. Singapore, for example, has implemented a sustainable financing model for its public transport system, using a combination of public and private funds. The city has also developed advanced regulatory frameworks and implemented security and privacy technologies to protect citizens' data. This integrated approach has enabled Singapore to build an efficient and safe transport system, overcoming financial and technological challenges.

In Vancouver, Canada, the city has developed educational and awareness programs to encourage the adoption of sustainable mobility practices. Vancouver has actively involved citizens in transportation planning and implemented economic incentives to ease the transition. These efforts have helped overcome resistance to change and promoted a culture of sustainability in the community.

Oslo, Norway, has implemented traffic management systems and digital platforms to integrate different modes of transport and improve the efficiency of urban mobility. The city has also developed flexible regulatory frameworks to adapt to the rapid evolution of technologies. These efforts have enabled Oslo to create a cohesive and efficient mobility system, demonstrating how technological integration and adaptive regulation can overcome barriers and promote sustainable mobility.

Although the promotion of sustainable mobility faces several challenges, these can be overcome with proper planning, cooperation across sectors, and a proactive approach to education and regulation. Examples from cities such as Singapore, Vancouver and Oslo show that it is possible to overcome these barriers and create efficient and sustainable transport systems that benefit both the environment and the economy.

Future prospects and emerging opportunities

As sustainable mobility technologies and approaches continue to evolve, the future prospects for the economic impact of sustainable mobility are promising. The convergence of technological advances, innovative business models, inclusive and equitable approaches, and infrastructure innovation is shaping a future in which sustainable mobility will not only be beneficial to the environment, but also a significant driver of economic growth.

Technological and digital advances will play a crucial role in this development. The use of artificial intelligence (AI) and machine learning will improve traffic management, urban planning, and transportation efficiency. These advances will optimize transport routes, reduce congestion and improve the punctuality of public services, which will translate into cost savings and greater satisfaction for users. The expansion of the Internet of Things (IoT) will enable greater connectivity and optimization of transportation systems, improving efficiency and convenience. Connected sensors and smart devices will be able to collect and analyze data in real-time, facilitating more informed and faster decisions. In addition, the use of blockchain to secure transactions and data can improve transparency and reduce fraud in mobility services. This technology can ensure the integrity of financial transactions and the exchange of information, increasing trust in mobility systems.

Innovative business models are also critical to the future of sustainable mobility. The sharing economy can provide innovative business models for shared mobility services, reducing costs and improving accessibility. Carsharing, bikesharing, and ridesharing platforms can offer affordable and flexible transportation options, decreasing the need for private vehicle ownership. The development of sustainable financing models, such as green bonds and social impact funds, can ensure that mobility solutions have the necessary resources to be implemented and maintained. These financial instruments can attract investment towards sustainable mobility projects, ensuring their long-term viability and success.

Inclusive and equitable approaches are essential to ensure that all citizens benefit from sustainable mobility solutions. Inclusive design should ensure that mobility solutions are accessible to all, regardless of their socio-economic, geographical or demographic status. This involves considering the needs of people with disabilities, rural communities, and other vulnerable populations in the design and implementation of transportation infrastructure and services. Equitable policies must promote social and environmental justice, ensuring that all communities benefit from advances in sustainable mobility. These policies can include subsidies for public transportation in low-income areas, the promotion of electric vehicles in disadvantaged communities, and the creation of cycling infrastructure in all urban areas.

Innovation in infrastructure will also play a crucial role in the future of sustainable mobility. Developing multimodal infrastructure that integrates different modes of transport and facilitates the transition between them will improve efficiency and convenience. For example, interchange stations where passengers can easily switch between trains, buses, bike shares, and electric scooters can make transportation more seamless and accessible. Integrating renewables into transportation infrastructure, such as solar charging stations and energy storage systems, will reduce environmental impact and improve sustainability. These innovations will not only decrease carbon emissions, but also ensure that transportation infrastructure is resilient and able to withstand future demands.

The future prospects for sustainable mobility are encouraging and full of potential. Technological and digital advances, innovative business models, inclusive and equitable approaches, and infrastructure innovation are shaping a future where sustainable mobility is not only an environmental necessity, but also an economic opportunity. With a comprehensive and collaborative approach, cities and communities can create transportation systems that are efficient, accessible, and sustainable, generating significant benefits for all citizens and contributing to robust and equitable economic development.

Conclusion

Sustainable mobility has a significant economic impact and offers numerous benefits in the short and long term. From reducing operating costs to creating jobs and driving economic development, sustainable mobility solutions can generate savings, create opportunities and foster sustainable economic growth.

As technologies and approaches continue to evolve, the opportunities to advance sustainable mobility and maximize its economic impact are immense. With the right approach, we can build a transportation system that not only meets our mobility needs, but also drives economic growth, improves quality of life, and protects our planet for future generations.

This chapter has explored in depth the economic benefits of sustainable mobility, providing examples of success and effective strategies. In the following chapters, we will continue to explore the technologies and policies that are shaping the future of mobility, with the aim of providing a comprehensive and compelling understanding of how we can build a more sustainable and equitable future for all.

Chapter 9: Sustainable Mobility in Rural Areas

Sustainable mobility is not only a concern of urban areas; It is also vital for rural regions and disadvantaged communities. However, the challenges and mobility needs in these areas are different from those in densely populated cities. This chapter explores how sustainable mobility solutions can be adapted and implemented in rural regions and disadvantaged communities, addressing the specific barriers they face and highlighting examples of success from around the world.

Challenges specific to rural regions

Rural regions face a number of unique challenges in relation to sustainable mobility, which require tailored approaches and innovative solutions. These challenges include low population density, large distances between communities, limited access to services and resources, and socioeconomic inequalities that complicate the implementation of sustainable mobility solutions.

Low population density and long distances in rural areas make the provision of public transport services more expensive and less efficient. Dispersed communities make it difficult to create an economically viable transportation infrastructure. This translates into a greater reliance on the private car as the main, and sometimes only, viable means of transportation due to the lack of alternatives. The need to travel long distances to access basic services and employment opportunities reinforces this dependency, further complicating the adoption of sustainable mobility practices.

Limited access to services and resources is another significant challenge in rural regions. Charging infrastructure for electric vehicles (EVs) is scarce, making it difficult to adopt these vehicles in rural areas. Without proper charging stations, rural residents may find it impractical to switch to electric vehicles, perpetuating reliance on internal combustion vehicles. In addition, public transportation options in these areas are limited, with schedules and routes often not meeting the needs of residents. A lack of reliable public transportation can further isolate rural communities, limiting their access to essential services such as healthcare, education, and employment.

Socioeconomic inequalities also play an important role in mobility challenges in rural areas. Average incomes in these regions are typically lower, which can make it difficult to adopt more expensive sustainable transportation technologies, such as electric vehicles. Rural residents may not have the financial resources to invest in new mobility technologies, and a lack of adequate incentives can exacerbate this situation. In addition, rural regions often receive less investment in transport infrastructure, perpetuating inequalities in access to mobility. This lack of investment can result in poorly maintained roads and the absence of public transportation services, further limiting economic opportunities and quality of life for rural residents.

To address these challenges, it is necessary to develop tailored approaches and innovative solutions that respond to the specific needs of rural communities. One possible solution is the implementation of flexible public transport systems, such as on-demand buses, that can adapt to the mobility needs of dispersed communities. These systems can improve access to public transportation without incurring the high costs associated with fixed routes and schedules.

Expanding EV charging infrastructure in rural areas is another crucial solution. Investing in fast-charging stations in strategic locations can facilitate the adoption of electric vehicles in these regions, reducing dependence on fossil fuels and improving sustainability. Specific incentives for EV adoption in rural areas, such as subsidies and tax breaks, can make these technologies more accessible to lower-income residents.

In addition, it is essential to foster public-private partnerships to attract investment in transport infrastructure in rural regions. Public-private partnerships can mobilize the resources needed to improve roads and develop new transportation solutions. It is also important to involve rural communities in the transport planning process to ensure that the proposed solutions respond to their needs and priorities.

Rural regions face unique challenges in sustainable mobility due to low population density, long distances, limited access to services, and socioeconomic inequalities. However, with adapted approaches and innovative solutions, it is possible to overcome these obstacles and improve sustainable mobility in these areas. The implementation of flexible public transport systems, the expansion of charging infrastructure for electric vehicles, and collaboration between the public and private sectors can all contribute to creating a more equitable and sustainable mobility system for rural communities.

Solutions adapted for rural mobility

To address the challenges of mobility in rural regions, it is necessary to develop sustainable mobility solutions that are tailored to their specific characteristics and needs. Implementing these solutions can significantly improve access to transportation and reduce reliance on the private car.

Long-range electric vehicles (EVs) are critical for covering the large distances typical of rural areas. EVs with higher-capacity batteries and extended range allow rural residents to travel significant distances without worrying about a lack of charging stations. In addition, installing fast-charging stations on strategic routes and points of interest, such as community centers and bus stops, can facilitate EV adoption in rural regions. These charging stations must be well distributed to ensure that EV users can access them conveniently and quickly.

Flexible and on-demand public transport is another key solution to improve mobility in rural areas. On-demand public transport services, which adapt to the specific needs of users in real time, can improve the accessibility and efficiency of transport in these areas. These services allow residents to request transportation when they need it, rather than relying on fixed schedules that may not be convenient. In addition, the use of micro-routes and shared vehicles, such as minibuses and collective taxis, can provide a viable alternative to traditional public transport. These smaller, more flexible vehicles can operate on specific routes based on demand, offering a more personalized and efficient service.

Shared and community mobility can also play an important role in rural regions. Implementing carsharing and ridesharing programs can reduce reliance on the private car and offer a more economical and sustainable transportation option. These programs allow residents to carpool, reducing individual costs and decreasing the number of vehicles on the roads. In addition, community-based transport initiatives, such as transport cooperatives, can organise and manage shared mobility services tailored to local needs. These cooperatives can coordinate carpooling among community members, optimizing resources and improving access to transportation.

The integration of technology and digitalization is crucial for the effective implementation of these mobility solutions in rural areas. The development of digital applications and platforms that integrate different modes of transport and provide real-time information can improve transport planning and access. These platforms can help users plan their trips, learn about available transportation options, and access on-demand transportation services. In addition, improving digital connectivity in rural regions is essential for the effective functioning of these platforms and services. Improved connectivity ensures that residents can use mobility apps and access crucial information at any time.

Addressing mobility challenges in rural regions requires tailored and specific solutions that consider the unique characteristics of these areas. Long-range electric vehicles, flexible and on-demand public transport, shared and community mobility, and the integration of technology and digitalisation are key approaches to improving transport accessibility and efficiency in rural regions. With proper implementation, these solutions can transform rural mobility, reducing reliance on private cars and promoting a more sustainable and accessible transport system.

Support policies and programmes

Public policies and support programmes are key to promoting sustainable mobility in rural regions. The implementation of economic incentives, financing programs, regulations, and public-private partnerships can significantly transform mobility in these areas, making it more accessible, efficient, and sustainable.

Economic incentives and subsidies are crucial tools to make sustainable mobility more affordable for rural residents. Subsidies and tax breaks for the purchase of electric vehicles (EVs) can reduce the initial cost of these vehicles, making them more accessible to people living in rural areas. This not only incentivizes the adoption of cleaner technologies, but also contributes to reducing carbon emissions in these regions. In addition, incentives for the installation of charging stations in rural areas can stimulate the necessary investments and facilitate the adoption of EVs. The availability of adequate charging infrastructure is essential for residents to consider EVs as a viable option.

Funding and grant programs also play a vital role in developing transportation solutions in rural areas. Grants for the development and operation of flexible, on-demand public transit services can help establish transportation systems that are tailored to the specific needs of rural residents. These programs can provide the resources needed to implement transportation services that are efficient and accessible. In addition, funds for community transport initiatives can empower local communities to develop and manage their own mobility solutions. Communities can use these funds to create transport cooperatives and other projects that directly respond to their local needs.

Regulations and standards are essential to ensure that mobility solutions are inclusive and equitable. Regulations that set accessibility standards for public transport and charging infrastructure can ensure that all people, regardless of their physical abilities, have access to transport options. This is particularly important in rural areas, where transportation options may be limited. In addition, transport integration policies that promote collaboration between different modes of transport can improve the efficiency and effectiveness of rural mobility solutions. These policies can make it easier to transition between different types of transportation, such as buses, bike shares, and electric vehicles.

Public-private partnerships are critical to the development and implementation of innovative mobility solutions in rural areas. Partnerships with technology and mobility companies can facilitate the development of solutions that are tailored to the specific needs of rural regions. These companies can provide technical expertise and resources that complement the efforts of the public sector. Cooperation with community-based organisations is also crucial to ensure that mobility solutions are relevant and sustainable in the long term. Community-based organizations can provide an in-depth understanding of local needs and help design and implement projects that are accepted by the community.

Public policies and support programmes are essential to promote sustainable mobility in rural regions. Economic incentives, financing programs, inclusive regulations, and public-private partnerships can transform mobility in these areas, making it more accessible, efficient, and sustainable. With a comprehensive and collaborative approach, rural regions can overcome the unique challenges they face and develop transportation systems that improve the quality of life for their residents.

Success stories in rural mobility

Several case studies from different parts of the world demonstrate how sustainable mobility solutions can be successfully adapted and implemented in rural regions and disadvantaged communities, showing a positive impact on the accessibility and sustainability of transport.

In Finland, rural areas face significant challenges in terms of access to public transport due to low population density and long distances. To address these issues, the Finnish government has implemented on-demand public transport services that use mobile apps to connect users with vehicles available in their area. These services are adapted to the specific needs of users in real time, significantly improving the accessibility of public transport in rural areas. This solution has reduced reliance on the private car and increased the efficiency of the transport system, providing a viable and flexible alternative for rural residents.

In Scotland, a lack of transport options and reliance on private cars are major challenges in rural areas. The community organization "Rural Car Club" has developed a carsharing program that allows residents to share vehicles through a digital platform. Vehicles are available at strategic points and can be booked in advance. This program has reduced the need for private vehicle ownership, lowering transportation costs and carbon emissions. In addition, it has improved transportation accessibility for residents who cannot afford a car of their own, fostering a sense of community and collaboration.

In New York's Hudson Valley, a region characterized by large distances between communities and limited access to electric vehicle (EV) charging infrastructure, a comprehensive solution has been implemented to encourage EV adoption. The local government, in collaboration with technology companies and community organizations, has installed fast-charging stations at strategic points and provided incentives for EV purchases. In addition, education and awareness programs have been developed to encourage EV adoption. These efforts have led to a significant increase in EV adoption in the region, reducing emissions and improving air quality. Fast charging stations have made everyday EV use easier, making them a viable option for rural residents.

In northern Australia, Indigenous and rural communities face significant challenges in terms of accessibility and mobility due to long distances and lack of infrastructure. To address these challenges, community transport cooperatives have been established that provide flexible transport services tailored to local needs. These services use minibuses and shared vehicles, and are managed by local communities. Cooperatives have significantly improved transportation accessibility, allowing residents to access essential services and economic opportunities. Community management ensures that services are relevant and sustainable, continuously adapting to the changing needs of communities.

These examples highlight how sustainable mobility solutions can be effectively adapted and implemented in rural regions and disadvantaged communities. Through innovation, collaboration, and community engagement, these initiatives have improved accessibility, reduced emissions, and fostered a greater sense of community. The key to success lies in adapting solutions to the specific characteristics and needs of each region, ensuring that they are sustainable and beneficial in the long term.

Implementation strategies and best practices

To ensure the success of sustainable mobility solutions in rural regions and disadvantaged communities, it is crucial to follow implementation strategies and best practices that are adapted to local needs and encourage community participation.

A local needs assessment is the essential first step. Conducting detailed assessments helps to understand the specific characteristics and challenges of each rural region and disadvantaged community. This assessment should consider factors such as population density, distances between communities, and existing infrastructure. Involving local communities in this process is critical. Community participation in planning and decision-making ensures that mobility solutions respond appropriately to residents' needs and preferences. When communities have a say in the design of these solutions, they are more likely to actively adopt and support them.

Capacity building and training are other key elements. Providing training and resources to transport managers and community leaders ensures that they have the skills and knowledge needed to implement and manage sustainable mobility solutions. This training should include technical and management aspects, as well as strategies to promote sustainable mobility among residents. In parallel, developing educational and awareness programs can foster a culture of sustainability. These programmes can include information on the environmental and economic benefits of sustainable mobility, as well as practical advice for adopting these practices in daily life.

Collaboration and partnerships are also crucial to the success of these initiatives. Fostering public-private partnerships allows for the sharing of resources and knowledge, and can be an important source of financing for the development and implementation of sustainable mobility solutions. Companies can bring innovation and technical expertise, while the public sector can offer regulatory and financial support. In addition, promoting inter-community cooperation is vital. Sharing best practices and developing joint solutions with other rural and disadvantaged communities can speed up the implementation process and increase the effectiveness of the solutions adopted.

Monitoring and evaluation are essential to ensure that sustainable mobility solutions are effective and stay relevant over time. Implementing continuous monitoring systems allows you to evaluate the performance and impact of these solutions, identifying areas for improvement and adjusting strategies as necessary. Regular impact assessments are equally important. These assessments should measure the social, economic and environmental benefits of mobility solutions, providing valuable data that can inform future decisions and policies. This evidence-based approach helps ensure that sustainable mobility initiatives are not only effective, but also sustainable in the long term.

The success of sustainable mobility solutions in rural regions and disadvantaged communities depends on careful planning, community engagement, capacity building, effective collaboration, and rigorous monitoring and evaluation. By following these best practices, it is possible to develop transportation systems that improve residents' quality of life, reduce carbon emissions, and promote greater equity in access to mobility.

Future prospects and emerging opportunities

As sustainable mobility technologies and approaches continue to evolve, the future prospects for rural regions and disadvantaged communities are promising. The adoption of emerging technologies, innovative business models, inclusive and equitable approaches, and infrastructure innovation are key elements to transform mobility in these areas.

The use of emerging technologies is at the heart of this transformation. Transport drones, for example, can be used to transport essential goods and services in hard-to-reach areas. This technology has the potential to significantly improve connectivity and reduce travel times, facilitating access to critical goods and services. Autonomous vehicles also offer new opportunities for on-demand transport and shared mobility services in rural regions. These vehicles can operate without the need for a human driver, which can reduce operating costs and increase transportation accessibility in areas where traditional services are scarce.

Innovative business models also play a crucial role in sustainable mobility. The sharing economy, which includes services such as carsharing and ridesharing, can provide shared mobility solutions that reduce costs and improve accessibility. These models allow users to share resources and services, which can be particularly beneficial in areas with low population density. In addition, the development of sustainable financing models, such as green bonds and social impact funds, can ensure that mobility solutions have the necessary resources to be implemented and maintained. These financial instruments can attract investment towards sustainable mobility projects, providing the necessary funds for their long-term development and operation.

Inclusive and equitable approaches are essential to ensure that mobility solutions benefit all citizens. Inclusive design involves creating mobility solutions that are accessible to all, regardless of their socio-economic, geographical or demographic status. This may include implementing disability-friendly transportation services and providing affordable transportation options for low-income communities. In addition, developing equitable policies that promote social and environmental justice is crucial. These policies should ensure that all communities, including disadvantaged communities, benefit from sustainable mobility solutions and have equitable access to transport services.

Innovation in infrastructure is another vital component for the success of sustainable mobility in rural regions. Developing multimodal infrastructures that integrate different modes of transport can facilitate the transition between them, improving the efficiency and convenience of the transport system. This can include creating transportation hubs where users can easily switch from one mode of transportation to another, such as from bike to bus or electric vehicle to train. In addition, the integration of renewable energies into transport infrastructure is essential to reduce environmental impact and improve sustainability. This can be achieved by installing solar charging stations and energy storage systems, which provide clean, renewable energy for electric vehicles and other transportation technologies.

The future prospects for sustainable mobility in rural regions and disadvantaged communities are encouraging, thanks to the adoption of emerging technologies, innovative business models, inclusive and equitable approaches, and infrastructure innovation. These strategies can transform mobility in these areas, improving accessibility, reducing carbon emissions, and promoting greater equity and sustainability in access to transport services. With a coordinated approach focused on local needs, it is possible to create mobility systems that benefit all citizens and contribute to the sustainable development of rural and disadvantaged communities.

Conclusion

Sustainable mobility in rural regions and disadvantaged communities presents unique challenges, but it also offers significant opportunities to improve quality of life, reduce environmental impact, and foster economic development. Through tailored solutions, supportive policies, community engagement and collaboration, it is possible to overcome these barriers and move towards a more inclusive and sustainable mobility future.

The examples presented in this chapter demonstrate that, with the right approach, it is possible to implement sustainable mobility solutions in various conditions and contexts. As technologies and approaches continue to evolve, rural regions and disadvantaged communities can benefit from emerging opportunities, ensuring that no one is left behind in the transition to a cleaner, more efficient and equitable transport system.

This chapter has explored strategies and examples of success in promoting sustainable mobility in rural regions and disadvantaged communities, providing a comprehensive and optimistic view of how we can make progress in this crucial area. In the following chapters, we will continue to explore the technologies and policies that are shaping the future of mobility, with the aim of providing a comprehensive and compelling understanding of how we can build a more sustainable and equitable future for all.

Conclusion: Towards a Sustainable Mobility Future

Throughout this book, we have explored various aspects of sustainable mobility, from technological advances and public policies, to citizen participation, education, and economic impact. We have analysed how these elements are interrelated and contribute to the transformation of our transport systems towards more sustainable, efficient and equitable models. In this conclusion, we will synthesize the key learnings from each chapter, highlight emerging challenges and opportunities, and offer a comprehensive and optimistic vision about the future of sustainable mobility.

Recap of key learnings

Technological advances are essential for the transition to sustainable mobility. Innovations in electric vehicles (EVs), charging infrastructure, telematics, and alternative fuels have significantly improved the efficiency, accessibility, and sustainability of our transportation systems. Next-generation electric vehicles, with improvements in batteries and powertrains, have increased range and reduced costs, making EVs a viable and attractive option for a growing number of consumers. Fast charging infrastructure and wireless charging technologies have improved the convenience and accessibility of EVs, facilitating their mass adoption. Fleet management systems and autonomous driving are transforming operational efficiency and transportation safety. In addition, biofuels, compressed natural gas (CNG) and hydrogen are promising options for reducing emissions in heavy-duty and long-range transport. Hybridization technologies and advanced exhaust systems are also helping to reduce emissions from internal combustion vehicles.

Smart cities are at the forefront of sustainable mobility, using advanced technologies and innovative urban planning approaches to improve the quality of life of their inhabitants and reduce environmental impact. Digital infrastructure and sensors, such as 5G communication networks and open data platforms, are essential for efficient traffic management and the optimization of mobility services. The electrification of public transport and the integration of transport modes into mobility-as-a-service (MaaS) platforms are improving the efficiency and accessibility of urban transport. Carsharing, bikesharing and ridesharing services, along with autonomous vehicles, offer flexible and sustainable transport solutions, reducing reliance on private cars. Sustainable urban planning, with low-emission zones, transit-oriented development (TOD), and cycling and pedestrian infrastructure, is encouraging active modes of transport and reducing congestion and emissions.

Citizen participation and appropriate education are essential to foster a culture of sustainability and ensure the adoption of sustainable mobility solutions. Public awareness and education campaigns, through the media, community events, and educational programs in schools, are effective in informing and motivating citizens to adopt sustainable practices. Public consultation, surveys and community committees allow citizens to voice their needs and concerns, ensuring that mobility solutions are inclusive and representative. Community ambassador workshops, courses, and programs provide the knowledge and skills needed to implement and maintain sustainable mobility solutions.

Sustainable mobility offers numerous economic benefits, from reducing operating costs to creating jobs and boosting economic development. EVs and sustainable modes of transportation, such as cycling and public transportation, have significantly lower fuel and maintenance costs than internal combustion vehicles. Route optimization and congestion reduction also improve operational efficiency. The transition to sustainable mobility is generating jobs in EV and battery manufacturing, charging infrastructure development, and shared mobility services. Investment in R+D and the startup ecosystem are also driving economic growth. Reducing air pollution and promoting active modes of transport improve public health, generating savings in medical costs and improving the quality of life of citizens. Subsidies, tax breaks and green finance programmes are facilitating the adoption of sustainable mobility solutions and ensuring a long-term return on investment.

Rural regions and disadvantaged communities face specific challenges in relation to sustainable mobility, but they also offer significant opportunities to improve quality of life and foster economic development. Low population density, long distances, and limited access to services and resources are major challenges in rural areas. Socio-economic inequalities can also make it difficult to adopt sustainable transport technologies. Long-range EVs, on-demand public transport services and community-based shared mobility initiatives are viable solutions for rural regions. Technology integration and digitalization can also improve transportation planning and access. Economic incentives, financing programmes and public-private partnerships are essential to foster sustainable mobility in rural and disadvantaged regions.

Case studies from different parts of the world demonstrate how sustainable mobility solutions can be successfully adapted and implemented in rural regions and disadvantaged communities. In Finland, on-demand public transport services use mobile apps to connect users with vehicles available in their area, significantly improving the accessibility of public transport in rural areas. In Scotland, the community organisation "Rural Car Club" has developed a carsharing programme that allows residents to share vehicles via a digital platform, reducing the need for private vehicle ownership. In New York's Hudson Valley, the installation of fast-charging stations and incentives for EV purchases have facilitated the adoption of these vehicles, improving air quality. In northern Australia, community transport cooperatives provide flexible transport services tailored to local needs, significantly improving transport accessibility.

To ensure the success of sustainable mobility solutions in rural regions and disadvantaged communities, it is crucial to follow implementation strategies and best practices. Conducting needs assessments to understand the specific characteristics and challenges of each region and community, and ensuring community participation in planning and decision-making, are essential steps. Providing training and resources for transport managers and community leaders, and developing educational and awareness programmes, can foster a culture of sustainability. Fostering public-private partnerships and cooperation between different communities can share resources and knowledge. Implementing continuous monitoring systems and conducting regular impact assessments can measure the social, economic, and environmental benefits of mobility solutions, and inform future decisions and policies.

As sustainable mobility technologies and approaches continue to evolve, the future prospects for rural regions and disadvantaged communities are promising. The use of drones to transport essential goods and services in hard-to-reach areas can improve connectivity and reduce travel times. Autonomous vehicles may offer new opportunities for on-demand transportation and shared mobility services. The sharing economy can provide innovative business models for shared mobility services, reducing costs and improving accessibility. Developing sustainable financing models, such as green bonds and social impact funds, can ensure that mobility solutions have the necessary resources. Ensuring that mobility solutions are inclusive and accessible to all citizens, and developing equitable policies that promote social and environmental justice are essential. Developing multimodal infrastructure that integrates different modes of transport and facilitates the transition between them, and the integration of renewable energy into transport infrastructure, can improve efficiency and sustainability.

Emerging challenges and opportunities

Despite the advances and benefits of sustainable mobility, there are still challenges that need to be addressed to ensure its long-term success. One of the main challenges is the initial cost and financing. Developing sustainable mobility infrastructure requires significant investments, which can be a barrier for many cities and communities. To overcome these barriers, it is crucial to implement sustainable financing models, such as green bonds and public-private partnerships. In addition, it is essential to ensure that all communities, including disadvantaged ones, have access to finance to develop sustainable mobility solutions, thus ensuring equity in their implementation.

Another major challenge is public acceptance and behavior change. Resistance to change and lack of interest can hinder the adoption of new technologies and sustainable transport modes. To encourage changes in behaviour, awareness campaigns and educational programmes are essential. In addition, it is vital to ensure that sustainable mobility solutions are inclusive and accessible to all citizens, regardless of their socio-economic or geographical situation.

The integration of technologies and regulations also presents significant challenges. Ensuring that different systems and technologies can work together efficiently is crucial to the success of sustainable mobility. International cooperation and standardization of technologies are essential to achieve interoperability. In addition, protecting citizens' data and ensuring the security of smart systems is critical to gaining public trust. Regulations must be flexible and adapt to the rapid evolution of technologies.

Finally, urban planning and sustainable development are essential for the success of sustainable mobility. Transit-oriented urban planning (TOD) focuses on development around public transport nodes, encouraging the use of public transport and reducing dependence on the car. The expansion of low-emission zones and pedestrian areas can improve air quality and promote active modes of transport, contributing to more sustainable urban development.

Despite advances in sustainable mobility, several challenges need to be addressed to ensure its long-term success. These challenges include upfront cost and financing, public acceptance and behavior change, integration of technologies and regulations, and urban planning and sustainable development. With concerted efforts in these areas, it is possible to create a sustainable mobility system that benefits all communities and contributes to a cleaner and healthier future.

Future prospects for sustainable mobility

The future of sustainable mobility is full of opportunities and challenges, and its success will depend on our ability to innovate, collaborate and commit to sustainability. One of the main drivers of this transformation will be technological and digital advances. The use of artificial intelligence (AI) and machine learning promises to improve traffic management, urban planning, and transportation efficiency, enabling continuous and adaptive optimization. The expansion of the Internet of Things (IoT) will enable greater connectivity and optimization of transportation systems, improving efficiency and convenience. In addition, the implementation of blockchain to secure transactions and data can increase transparency and reduce fraud in mobility services.

Innovative business models will also play a crucial role in the future of sustainable mobility. The sharing economy can provide new models for shared mobility services, reducing costs and improving accessibility. The development of sustainable financing models, such as green bonds and social impact funds, can ensure that mobility solutions have the necessary resources to be implemented and maintained. These financial innovations are essential to support the growth and expansion of sustainable infrastructure.

In terms of inclusive and equitable approaches, it is essential to ensure that mobility solutions are accessible to all citizens, regardless of their socio-economic, geographical or demographic situation. The inclusive design of transport infrastructure and services is key to ensuring that no one is left behind. In addition, developing equitable policies that promote social and environmental justice is essential to ensure that all communities benefit from sustainable mobility solutions.

Innovation in infrastructure is another crucial area for the success of sustainable mobility. Developing multimodal infrastructure that integrates different modes of transport and facilitates the transition between them will improve the efficiency and convenience of the transport system. In addition, integrating renewables into transportation infrastructure, such as solar charging stations and energy storage systems, can reduce environmental impact and improve long-term sustainability.

The future of sustainable mobility depends on our ability to harness technological and digital advances, develop innovative business models, implement inclusive and equitable approaches, and foster infrastructure innovation. With collective commitment and effective collaboration, it is possible to create a transportation system that not only meets our mobility needs, but also promotes the sustainability and well-being of all communities.

Vision for a sustainable future

The transition to sustainable transport is a complex and multifaceted task that requires the cooperation and commitment of all sectors of society. Through technological innovation, citizen participation, education, and effective public policies, we can move toward a cleaner, more efficient, and equitable mobility future. To achieve this goal, it is essential that strong ties of collaboration and cooperation are established between governments, businesses and communities. Public-private partnerships, international cooperation, and community engagement can facilitate the development and implementation of sustainable transportation solutions.

Collaboration between different actors not only allows for the sharing of resources and knowledge, but also fosters innovation and creativity in the search for effective solutions. Technological innovation and the ability to adapt to changing market needs are essential to continue advancing sustainable mobility. Cities and businesses must be willing to try new ideas, learn from experience, and adjust their strategies as needed. This implies being open to experimentation and the implementation of new technologies, as well as to the continuous re-evaluation and adjustment of existing policies and practices.

A strong commitment to sustainability is critical to ensuring that future generations can enjoy a clean and healthy environment. Transport policies and strategies must prioritise emission reduction, energy efficiency and social equity. This means adopting practices that minimize environmental impact and promote social and economic well-being. Transportation decisions must be guided by sustainability principles, always seeking to balance development with the preservation of the environment and the well-being of communities.

The transition to sustainable transport is not an easy task, but it is an achievable goal if they are approached with a collaborative, innovative and committed approach to sustainability. It is crucial that all sectors of society work together to develop and implement mobility solutions that are clean, efficient and equitable. Through cooperation, adaptation to new technologies and a strong commitment to sustainability, we can build a future of mobility that benefits everyone.

Conclusion

Sustainable mobility is not only a desirable goal; It is an urgent need to meet the environmental, social and economic challenges of our time. In the pages of this book, we have explored how advanced technologies, effective public policies, active citizen participation, and economic benefits can converge to create a more sustainable and equitable transportation system.

Climate change, air pollution and urban congestion are problems we can no longer ignore. The decisions we make today will determine the world in which our future generations will live. Sustainable mobility offers a powerful and viable solution to these problems, providing a form of transport that respects the environment, fosters social inclusion, and supports economic growth.

We've seen how advances in electric vehicles, charging infrastructure, and telematics technologies can transform our cities and communities. Electric vehicles significantly reduce greenhouse gas emissions and air pollution, while fast-charging stations and wireless charging solutions make it more convenient and accessible for everyone to adopt this technology. In addition, telematics and intelligent traffic management systems improve operational efficiency and road safety.

Public policies play a crucial role in this process. Governments have a responsibility to create an enabling environment for sustainable mobility by implementing regulations that promote the use of clean technologies, investment in infrastructure, and offering economic incentives. Subsidies for the purchase of electric vehicles, tax exemptions for the installation of charging stations, and investments in sustainable public transport are just some of the measures that can boost the adoption of sustainable mobility practices.

Citizen participation is also essential. Citizens must not only be informed about the benefits of sustainable mobility, but they must also be involved in the planning and decision-making process. Public consultation, surveys, and the formation of community committees are powerful tools to ensure that mobility solutions reflect the needs and preferences of the community. In addition, education and awareness can help change behaviors and foster a culture of sustainability.

The economic benefits of sustainable mobility are significant and far-reaching. Reducing operating costs for businesses and individuals, creating jobs in new tech industries, and improving public health are just a few of the benefits we've discussed. Cities and communities that invest in sustainable mobility are not only protecting the environment, but they are also creating a strong foundation for sustainable and equitable economic growth.

The success of sustainable mobility depends on our ability to work together, innovate and maintain a strong commitment to sustainability. This requires a shared vision and a collective effort to overcome the challenges and seize the opportunities that present themselves. With the right approach, we can build a transportation system that not only meets our mobility needs, but also protects our planet and improves the quality of life for all.

We hope that the lessons and strategies presented in this book will inspire governments, businesses and citizens to take concrete action and contribute to the creation of a cleaner, more efficient and equitable mobility future for future generations. The transition to sustainable mobility will not be easy, but with determination and collaboration, we can move towards a future that benefits everyone and ensures a healthier and more prosperous planet for all.

Let's imagine a world where our cities are free of air pollution, where transportation is accessible to all, and where our communities are vibrant and sustainable. This is the future we can build if we work together and commit to sustainable mobility. Every action counts, every decision matters. Together, we can make this future a reality.

www.ingramcontent.com/pod-product-compliance
Lightning Source LLC
Chambersburg PA
CBHW052154220526
45471CB00004B/1672